UNDERSTANDING SCIENCE THROUGH EVOLUTION

Publication Number 1069
AMERICAN LECTURE SERIES®

A Publication in

The BANNERSTONE DIVISION *of*
AMERICAN LECTURES IN ENVIRONMENTAL STUDIES

Consulting Editor

CHARLES G. WILBER, Ph. D.

Professor of Zoology
Colorado State University
Fort Collins, Colorado

UNDERSTANDING SCIENCE THROUGH EVOLUTION

A Humanist Approach

By

ARNOLD M. CLARK, PH.D.

Professor of Biology Emeritus
University of Delaware
Library Reader
Marine Biological Laboratory
Woods Hole, Massachusetts

CHARLES C THOMAS • PUBLISHER
Springfield, Illinois U. S. A.

Published and Distributed Throughout the World by

CHARLES C THOMAS • PUBLISHER
2600 South First Street
Springfield, Illinois 62794-9265

ISBN 0-398-05312-X

Library of Congress Catalog Card Number: 86-30023

With THOMAS BOOKS *careful attention is given to all details of manufacturing and design. It is the Publisher's desire to present books that are satisfactory as to their physical qualities and artistic possibilities and appropriate for their particular use.* THOMAS BOOKS *will be true to those laws of quality that assure a good name and good will.*

Printed in the United States of America
Q-R-3

Library of Congress Cataloging in Publication Data

Clark, Arnold M.
Understanding science through evolution.

(American lecture series ; publication no. 1069)
Bibliography: p.
Includes index.
1. Evolution. 2. Evolution--Religious aspects.
I. Title. II. Series.
QH371.C58 1987 575 86-30023
ISBN 0-398-05312-X

To Connie, Dan, Jane, Rags

and

Their Common Ancestors

FOREWORD

THE CONCEPT OF evolution is the unifying framework for contemporary biology. The concept is derived from a deluge of experimental and observational data confirmed and reconfirmed many times over. The evolutionary idea is truly one of the few valid, scientific generalizations which are of real concern to all liberally educated persons.

What then are we to say about the strife involving "evolution," "creationism," "atheistic science," and related terms and phrases that have even moved into our civil litigation machinery?

In this present volume Arnold Clark draws upon his decades of teaching, writing, and lecturing on evolution as one of the liberal studies to be pondered seriously by all educated persons, whether scientists or not. He has woven together a well-written, readily understood book which should interest professional biologists, students, and other liberally educated persons such as clergymen, teachers, educators, and students of ideas.

Much confusion surrounds the idea of evolution in the minds of the general public. It is important to emphasize that evolution as a scientific generalization is virtually unanimously accepted among biologists the world over as a fundamental scientific scheme for understanding the world of living things. Indeed, such biologists as the late Fr. Pierre Teilhard de Chardin have argued that evolution is a FACT of biology and that to understand all physical reality one must view it through the lens of the evolutionary consensus.

The concept is attacked primarily by a comparatively small number of self-named "creationists" or "creation scientists" who would use the Bible in literal terms as a scientific outline of how the universe began and how life came to be what it is on our planet.

The general viewpoint about biology and the Bible held by most liberally educated persons of Judeo-Christian persuasion may be fairly expressed as follows: the Bible is not in any sense a textbook about anything. The Bible is by no stretch of imagination a guide or outline of how life originated and developed physically on earth. The scientific method reveals material truths which can be tested by experiment and which have a validity that cannot be "second-guessed" by religious argumentation.

The study of evolution, as Doctor Clark so aptly outlines, opens up to the liberal student the respective roles of genetic makeup and environmental factors in shaping not only individual characteristics but also the specific details of historical life on environmental ethics, quality of life, and similar ideas that will be with us for a long time in the future. It is mandatory for educated persons to understand how life, including human life, as we know it today came to be. Knowledge of the process by which environmental variables (accidents?) molded the expression of genetic factors during the history of life on earth should aid in a clearer understanding of contemporary life processes and perhaps will lead to better and more positive decisions with respect to environmental variables that mankind can control.

Often, the poet or the *litterateur* expresses more clearly a fundamental concept about nature than does the scientist. Hilaire Belloc, who had a lifelong love affair with the ocean, eloquently spoke of the ocean as our evolutionary womb: "There, on the sea, is man nearest to his own making, and in communion with that from which he came, and to which he shall return. For the wise men of very long ago have said, and it is true, that out of the salt water all things came. The sea is the matrix of creation, and we have the memory of it in our blood." Truly human, humane studies ought to include a pursuit of that memory by the use of all the tools available to modern science and scholarship.

Charles G. Wilber, Ph.D.
Fort Collins, Colorado
Fall, 1986

PREFACE

THE FUNCTION of basic science is to describe nature and discover the laws that govern nature. This knowledge often leads to a technology whose function it is to manipulate nature for the benefit of mankind. This technology may have an impact on the economic or political aspects of society. The discovery of radio waves by Helmholtz led to the development of the wireless by Marconi and to the formation of an important communications system. The discovery by Einstein that energy and matter can be interconverted led to the development of nuclear energy and the atomic bomb. The discovery of microorganisms by Pasteur and others led to the control of infectious disease. Certain scientific ideas, however, may have an impact on moral values, on how we view ourselves and our importance in the universe. The idea of the rotation of the earth around the sun put forth by Copernicus and Galileo questioned that the earth was the center of the universe. The idea of gravitation put forth by Newton gave a natural, in contrast to a divine, explanation for the orderly rotation of the planets. The idea of evolution put forth by Darwin questioned the biblical views of creation.

It takes many years before an idea with moral implications for one generation becomes a part of the accepted truth of another. This appearance of new ideas is the story of the intellectural history of man, and every generation is challenged to adapt. This development and assimilation of ideas needs to be viewed in historic perspective. The idea of biological evolution confronts us today just as planetary rotation challenged us yesterday. We need to appreciate that it will not bother us tomorrow but that some other ideas will be "troublesome and immoral."

The concept of biological evolution is one of the big ideas in Western civilization. It has been important in shaping our attitudes and insights into the world around us and within ourselves. What makes it a big idea? It is a big idea because it illuminates so many fields of thought. The concept of biological evolution is one of the major ideas upon which modern-day biology is built. It is crucial to the sciences of geology, geography, astronomy, and astrophysics. It is important in the field of behavior and emotions. It bears upon racial similarities and differences, competition, and cooperation among groups. It has had an impact on philosophy and religion by its challenge to the Judeo-Christian biblical view of creation. It bears upon the meaning of purpose in our origin and in our lives.

In this book I will try to present the subject of biological evolution to teachers, students and other citizens. These are people who need to be interested not only in its biological but also its social and humanistic aspects. I plan to treat evolution as a science-humanities experience, one that not only considers the sciences but also inquires into our beliefs, attitudes, and the nature of evidence. Science and religion represent two of the ways of getting at truth. What is the difference in approach between them? In a modern world where so many important and unanticipated scientific discoveries are being made, what is one supposed to think about people and their importance? Every person in every age needs to confront himself with the goals of and the differences between science and religion, and the nature of evidence.

The present controversy in the United States relating to the teaching of creationism and evolution to public school students needs to be taken as a challenge to a teacher. It is the job of the teacher to turn this controversy into a learning experience for students so that they can begin to discern the nature of evidence and to sort out for themselves the methods and purposes of science and religion. They need to "render unto Caesar that which is Caesar's and to God that which is God's."

I am sometimes asked, "Do you believe in evolution?" My answer is, "Are you asking a scientific question or a religious question?" "What's the difference?" he asks. A belief may be based on evidence or it may be based on faith. From the standpoint of science, the evidence has convinced most biologists that evolution has occurred and is occurring by means of natural laws. Millions of people in the United States believe, however, that the world and the species upon it were created by God in ways that transcend natural law. For them the concept of evolution is a challenge to biblical literalism and to the concept of a personal God. Before people can examine the scientific evidence for evolution, they will need to examine their faith. The accommodation of revelation and reason is a major humanistic conflict to face people who are living in a scientific age. This conflict needs to be resolved before one can meaningfully read the many erudite texts dealing with biological evolution. In a learning situation one needs to address the religious and emotional aspects in order to appreciate the scientific aspects. Indeed, the comparison cannot honestly be avoided.

During the 35 years that I was a professor at the University of Delaware, I was heavily involved with the teaching of biology to future biologists in order to help them to learn the concepts and techniques to achieve their professional goals. I also taught courses for non-majors in order to present biology as a liberal arts subject and to try to relate the sciences to the humanities. It became increasingly clear to me that one cannot teach evolution in the same way that one teaches respiration or mitosis. Biological evolution is thought by many people to be primarily a religious matter and a moral issue. Presentation of evidence accumulated by scientists is, therefore, of little consequence. The biology teacher who presents evolution may think that he is teaching science, but the student-citizen may think that he is teaching a religion and destroying traditional faith.

The present essays have been written from my teaching and learning experience. It is an attempt to bring out the issues that have bothered people. I hope that it will bridge a gap between the esoteric erudition of scientists and the religious morality of citizens. Evolution needs to be treated as a case study in the understanding of science (and religion). I hope that you will approach this book with an attitude of thoughtfulness and open-mindedness. This book will require you to search your opinions through questions that are asked. Its reading will require active participation.

It is often difficult to approach objectively those subjects that have religious and social consequences. Throughout this book I have expressed opinions for your reaction. I am not, however, giving you "the word." I will be successful not by convincing others of my opinions, but rather by stimulating thought and discussion. In a very real sense, education is frequently an invasion of privacy. It is poignant that only in the classroom is evolution studied merely as biology. In the non-academic world, evolution is treated as history, philosophy, religion, poetry, politics, and education. Evolution is an important scientific and humanistic concept. It should, therefore, be studied not only as science but also as intellectual history. Evolution is an interesting idea. Let's play with it.

ACKNOWLEDGMENTS

I WISH TO THANK the people who have reviewed all or part of my manuscript and have made suggestions. Useful discussions were held with colleagues at the University of Delaware: Adair Gould, Rivers Singleton, David Sheppard, Steven Skopik, Harold White, Jay Halio. I wish to apologize to Rivers for the many times that I have cornered him in his office to hold a discussion. Joyce Shaffer and Carol Horgan were helpful for their views as social science teachers; the Rev. Kenneth Smith and the Rev. William Clark for their assessment of the manuscript from the viewpoint of the clergyman; Cornelia Cummings and Fred Pratson for their publishing suggestions; Jane Townsend and Drusilla Harding for encouragement. Paul Gross, Director of the Marine Biological Laboratory, reviewed my manuscript and made very useful suggestions. My special thanks are given to the librarians at the Marine Biological Laboratory. They have furnished willing and intelligent support. I thank Jane Leighton for her conscientious and competent typing and Barry O'Neil for Figure 1. I am grateful to my editor Charles G. Wilber for his needed help. I wish to thank my good friend and colleague, Daniel S. Grosch, for many useful conversations.

CONTENTS

UNDERSTANDING SCIENCE THROUGH EVOLUTION

CHAPTER 1

A MATTER OF ATTITUDE

LAST YEAR, I met a former student whose daughter, Jane, was a sophomore in high school. "My daughter," she said, "was taking a biology course with Mr. Z who emphasized the concept of evolution. This upset Jane to such an extent that she changed to a biology course in which the teacher did not present the concept of evolution. What was particularly upsetting to Jane was that she had just finished a 10-week course in summer Bible school." My reply to the mother was, "I'm sorry." On the way home, however, I began to think concerning WHAT and HOW teachers present information in Bible school that so scares these students when they get to public school. The daughter was shocked and the mother was offended. What kind of Bible school is this? What kind of a parent is afraid of scientific ideas?

In 1980, in the State Supreme Court of California, a man objected to the teaching of evolution to his son who was studying biology in the public schools. He stated that he was personally offended. To him, evolution was a kind of religion. He petitioned the state of California through the law courts to prohibit evolution from being taught unless equal time were given to teach the view of creation that is presented in Genesis. Such a law for equal time was later passed by the legislatures of Arkansas and Louisiana and was challenged in the courts of law. The law was declared unconstitutional by the federal courts.

During the 1920s there was an effort by fundamentalistic groups to prevent the teaching of evolution in the public schools. The state of Tennessee passed an anti-evolution law and brought to trial John Scopes, a teacher who presented evolution in the classroom. Even though Scopes was not punished, there was very little teaching of evolution from 1925 to 1964 in the public schools in the United States. Textbooks were carefully censored and bore no mention of Darwin nor evolution. Even though there was less pressure at the state level, nevertheless, at the local level lots of pressure existed. What teacher would dare to present evolution? Today, we have a resurgence of creationism by fundamentalists who insist that both evolution and creationism be taught side by side with equal time and equal fervor in science classes. Already, textbooks have been rewritten for public schools in the United States to exclude mention of evolution. Creationism is an attack not only on biology and the concept of the origin of species but also upon geology and physics and the age of the earth, upon astronomy and the age of the universe. It is an attack on scientific methods and scientific thinking. It is also an attack on biblical scholarship and biblical interpretation. How is one to read the Bible, literally or metaphorically?

Thus, the concept of evolution, which was put forth to explain the endless variety and diversity of millions of species, is an idea that not only has an impact on the biological sciences but also on the social sciences and the humanities. Biological evolution has not only scientific but also religious implications. How we view the world around us is dependent upon both faith and reason. Faith is the bulwark of religion; reason is the bulwark of science. In most cases these two can be kept separated in our minds, but sometimes we run across an idea that seems to challenge both. Such an idea is evolution. What do we do when there is a difference of opinion based on religious versus scientific grounds? For all of us there is a world that is and a world that ought to be. In an age of science, what is one to believe about traditional religion? The concept of evolution is not the first time in human history that this question has been asked. Throughout the history of mankind the interaction of science and religion has persisted, most notably with respect to the Copernican Revolution. It is one of the major themes that runs through human history. Each generation and each individual must face this anew and must pass these opinions on to the next generation. What do we teach children and how should they be taught?

The biologist looks upon evolution as a characteristic of living things. He considers evolution as a science rather than as a religion. What does this mean? He tries to marshall the data (the evidence). He asks about observations that have been made that can be explained by the concept of evolution. He inquires concerning the experiments that have been done. He looks at the evidence and draws conclusions from this evidence. These conclusions in science are not discarded solely because they are deemed immoral or atheistic from a religious point of view or do not seem to conform to Scripture. He inquires into the limitations of his observations and conclusions. He knows that scientific concepts are frequently modified. There is a spirit in the scientific community to challenge and to question. No truth is untouchable. If the concept of evolution is false, it will be shown to be false by scientists working in their laboratories and collecting data. It will not be shown to be false by clergymen or philosophers.

THE NATURE OF EVIDENCE

Over the years there have been many ideas to which you were exposed. What is the basis for your acceptance or rejection of these ideas? Go to the dictionary and define "direct, circumstantial, hearsay evidence." Think out an example of each of these types of evidence. What is the basis for your acceptance or rejection of the following:

- There is a country called Kenya in Africa.
- Men have walked on the moon.
- The Philadelphia Phillies won the World Series Baseball Championship in 1980.
- The present economic difficulties of the United States are due to high interest rates.
- Communism is the best possible system of government.
- There is a God.
- There was once a time in the history of man when the techniques of agriculture were not known.

There are people who deny that the earth is a sphere. They consider space flights to be a hoax. What do you answer them? Should you answer them? We knew (or thought we knew) long before space flights that the earth was a sphere. Describe the evidence that the earth is (a sphere) (flat).

List some "facts" that you accept or reject because of personal bias.

List some "facts" that you accept or reject that require no personal bias.

Here are some concepts that were developed by the scientific method. Which of these concepts do you accept or reject? What is the basis (rational or irrational) for your acceptance or rejection of these concepts?

1. The earth rotates around the sun.
2. The moon is closer than the sun to the earth.

3. Genes are made of a chemical called DNA.
4. Blood circulates in veins and arteries and in one direction.
5. Light travels at a speed of 187,000 miles per second.
6. Green plants make their own food by means of photosynthesis.
7. The gas that is used to keep a candle burning is the same gas that a mouse needs to breathe.
8. Houseflies arise from other houseflies and not from dead decaying material.
9. The front flipper of a whale has a bone structure that is more similar to that of a man than to that of the pectoral fin of a fish.
10. The beagle, the Pomeranian, and the Great Dane all belong to the same species.

The most direct way to determine the diameter of the earth at the equator is to pull a long string through the earth and then measure it. Why has this not been done? Is it possible to determine indirectly the diameter of the earth at the equator? How can this be done? Are you satisfied with your indirect measurements?

If the earth moves around the sun, the alternating days and nights can logically be explained by assuming that the earth rotates on its axis. Galileo arrived at the conclusion that the earth does indeed rotate on its axis, but he could not prove it. This was proved experimentally by Foucault about 1850. What did Foucault do?

Describe an experiment to test the concept that "the gas that is used to keep a candle burning is the same gas that a mouse needs to breathe."

Go to the dictionary and look up the word "belief." On what basis do you believe the following:

1. I believe that the earth is a sphere.
2. I believe that there is a God.
3. I believe that there once existed a man called Julius Caesar.
4. I believe that democracy is the best system of government.
5. I believe that Halley's comet will appear in the year 2062.

John Jones has been brought to trial for the murder of his wife. No one saw the murder. On what basis does the judge and jury make a decision "without reasonable doubt?"

Faith

Faith can be defined as an unquestioned belief, one that does not require proof or evidence. In science there are no unquestioned beliefs. All beliefs are challenged and require proof or evidence. Much of traditional religion has been built upon unquestioned beliefs. There are other definitions of "faith." How do you define "faith?" How do you apply this definition to religion and to science?

THE ISSUES

What are the issues in the present evolution-creationism controversy? They are not whether one is right and the other is wrong. The idea that the earth was created about 6,000 years ago and that creation occurred over a period of six successive days is a testable and a thoroughly tested hypothesis. I would certainly accept this idea if it were proved by scientific methods. It is rather the fundamentalism of this idea with which I take issue. Creationism is tied up with morality and the supernatural. One must believe it out of fear for one's salvation. Were the idea of evolution to be forced upon me and confused with morality and atheism, and if I were told that I must believe the concept or suffer punishment, I would certainly be resentful of it. The reason that different religious groups migrated to the United States was to have religious freedom. Freedom of religion means freedom from the imposition and threat of other religious beliefs. Creationism is a religious belief that is being imposed upon public and secular schools. One cannot question; one must believe, or else. Evolution-creationism is an important educational issue, because it relates to that uniquely human right of each of us to question, to challenge, to innovate, and to think. The issue was stated admirably by Thomas Jefferson in 1800: "For I have sworn on the altar of God, eternal hostility against every form of tyranny over the mind of man."

CHAPTER 2

THE SPIRIT OF SCIENCE

I BELIEVE that science and the scientific method is a way to describe and explain the world around us. In considering the scientific method one is concerned primarily with an attitude that includes: (1) I may be wrong; (2) no truth is untouchable; (3) data must be explained by natural laws; and (4) conclusions must be based on observation and experimentation.

There is the chance that I may be wrong. There are things that I have taught that have turned out to be wrong: that the diploid number of chromosomes in man is 48; that genes are composed of proteins. Science proceeds by making mistakes. It gives a more accurate view of nature as it corrects its mistakes. Although there are absolute truths in the universe, these are not easy to perceive. Each generation describes its view of the world at a particular time and within a particular world view. There is the need to keep the channel open so that subsequent generations can improve upon our ideas or at least not be stifled by us. The people who follow me must be taught in such a way that they do not hesitate to challenge my ideas. I am convinced that the earth revolves around the sun, but I know that at one time many of our best minds were convinced that the sun goes around the earth and that the earth was the center of the universe. Science is a process of inquiry; it is not a body of unchangeable conclusions. We need to accept that we can challenge every idea, however well-established in the minds of people.

Scientific concepts may come about by intuition and insights shown by very bright people, but proof of the validity of a concept comes from many observations and experiments. One needs to marshall the data; clever arguments are not enough. In a scientific laboratory one is sometimes confronted with a good idea versus contrary data. Darwin spoke of "a beautiful theory destroyed by an ugly fact." I have a simple example of this from my own experiences.

About 40 years ago there were many research efforts dealing with the harmful effect of high energy radiations. It was discovered that when organisms were irradiated in the absence of oxygen, they were protected to a certain extent from radiation damage. This was called "the oxygen effect." I did an experiment to test the oxygen effect on insect development. I exposed pupae of a wasp *(Habrobracon)* to x-irradiation under the following conditions. White (young) pupae were placed into plastic chambers. The chambers were flushed with oxygen or nitrogen for thirty minutes, sealed, and exposed to 10,000 r. The following data in Table I show the results.

TABLE I

EFFECTS OF VARIOUS GASES ON RADIATION SENSITIVITY FOR PUPAE

X-ray dose (r)	*Gas*	*No. pupae treated*	*No. adults*	*Mortality %*
10,000	O_2	100	0	100
10,000	N_2	100	100	0
0	O_2	100	0	100
0	N_2	100	100	0

Examination of Table I shows that for the two groups of pupae exposed to 10,000 r, those pupae that were irradiated in the presence of oxygen failed to develop to the adult stage, while those pupae exposed to x-rays in nitrogen (absence of oxygen) continued their development to adults. These are the expected data. But one needs to look at the controls, the pupae that were exposed to the gases alone. The oxygen-treated pupae, without exposure to x-rays, showed the same amount of mortality as the oxygen-treated pupae that were exposed to x-rays! Thus, I was not measuring mortality due to x-rays but rather to oxygen. At the time (1952), the data confused me. I knew that wasps needed oxygen to survive (if kept in nitrogen, they die) and yet the data do not seem to show this. Question: "Is there something wrong with my idea that oxygen is not harmful but nitrogen is harmful, or did I goof in the design or performance of this experiment?" What should I do? What would you do? Well, I repeated the experiment and I obtained the same kind of data. In fact, I repeated the experiment five times, and after that I was convinced that I should discard my hypothesis that oxygen was harmless. I went to the library (I should have done this sooner) to see if someone had ever reported this finding, and I found that already in 1879, a Frenchman, Paul Bert, had made observations on the toxic effects of oxygen. Bert had discovered "oxygen poisoning." We need oxygen to survive, but oxygen can also be a toxic substance. Contradiction: "How can oxygen be toxic when we need it for survival; indeed, we would die almost instantly without it?" The oxygen poisoning experiment that I did, opened up a field of investigation that I pursued. And what I learned from the initial experiment was: do not be hasty to get rid of a bizarre idea; it may be true.

I believe that scientists are almost always honest in the reporting of data. They are, however, no more honest in their search for truth than clergy or lawyers or teachers. When a scientific paper is published, it may be badly designed or it may be badly interpreted, but the author presents what he considers to be an honest effort. He explains his experiment so that it can be repeated by others. If different laboratories do the experiment and get the same data, then the data are accepted and become part of the general knowledge. When scientists disagree on the data, they try to determine what were the conditions that caused different experiments to turn out differently. They often have heated arguments on how to interpret the same data. I accept the observations of my colleagues as honest. I cannot repeat all of the observations in biology. I need to rely on the conclusions of other scientists. Although I have not personally made the observations, I believe that genes are made of DNA, that mammals have a four-chambered heart, and that malaria is caused by a protozoan parasite. I accept the conclusion reached by my colleagues in geology and astronomy that the earth and the universe are billions of years old. There have been many honest mistakes in science. There have been very few hoaxes.

Scientists attempt to answer questions by natural explanations. They attempt to explain Down's syndrome, a disease that causes mental and physical disability, in terms of chromosomal non-disjunction and the presence of an extra chromosome no. 21. Down's syndrome is an observable and describable event and therefore lends itself to scientific explanation. When in 1959 a child with Down's syndrome was shown to have an extra chromosome-21, there was the possibility that this was the cause of the disease. One could set up the hypothesis: Down's syndrome is due to the inheritance of an extra chromosome-21, so that the fetus will have 3 rather than 2 chromosomes-21. This is an operational (testable) hypothesis. What one does is to examine the chromosomes of children who clinically show Down's syndrome and compare them with normal children. Based upon such studies it can be shown that trisomy-21 is a cause of Down's syndrome. And based upon this law (established hypothesis) one can predict that a fetus with trisomy-21 will be born with Down's syndrome. Based upon this law one can make use of amniocentesis to advise parents whether or not they will bear a child with an extra chromosome-21. One can explain Down's syndrome because of a previous event that occurred in the formation of a gamete, the event of non-disjunction. The cause of Down's syndrome is an extra chromosome-21. The cause of the non-disjunction is at present not known, but, since it is also an observable and describable event, it is within the realm of science to be described by a natural cause. When this is discovered, it may then be possible to control non-disjunction and thereby do away with the scourge of Down's syndrome.

One may suggest that Down's syndrome is caused by a sinful parent or by emanations from outer space or by some event that we dare not examine or question or for some future purpose to the glory of mankind, but these are not scientifically testable hypotheses. Science is limited in its scope, in that it can only search for natural explanations. It cannot cope with supernatural explanations. It is not designed to cope with supernatural explanations.

If one asks a biologist "What is life?" he cannot answer. The biologist will put the question in operational terms, "What are the characteristics of living things?" Questions such as "What is life?", "What is the purpose of human existence?", "Is there a God?" are addressed by philosophers. These latter are important questions, but they are not within the scope of science to answer.

One needs to approach the formulation of truth with a certain amount of humility. Each of us is "a child of his age." If I had been born 500 years ago, I would have believed that the sun rotates around the earth. If I were born in the Soviet Union 50 years ago, I would have denied the existence of genes. Each of us is limited by his environmental experiences (by his imprinting). This calls for the need for tolerance.

The scientific community has a spirit of optimism. It is used to addressing tough problems. It is confident that those problems that it cannot solve now will be solved in the future. It knows also that some questions concerning the laws of nature will arise in the twenty-first century of which we are at present completely unaware. Before the twentieth century, we were unaware of x-rays or radioactivity. Before the seventeenth century we were unaware of the circulation of blood. In science, we try to maintain an open mind regarding the natural laws that guide the universe.

The Method of Doubt

Rene Descartes (1596-1650) presented the method of doubt into scientific work: "We must doubt everything." This idea has been important in the development of science. Through the method of doubt, self-critical analysis, we get closer to the truth. Pierre Bayle (1647-1706) applied the method of doubt to history and to religion. He thought that the great danger to true religion was a religious dogmatism which claimed that the Bible and tradition were the true sources of historic fact. Bayle freed history from the chains of theology. His work led to a science of history that is based upon the concept that "historic facts" should not be accepted until they have first been examined. The scientific study of religion that developed in the nineteenth century is based upon the right to doubt the view of history presented in the Bible. The French philosophers in comparing customs and religions throughout the world found that some ideas that were thought to be ridiculous by some people are held as sacred by another. It is through doubt that science, history and religion can approach the truth. The responsibility to question leads to innovation and to change. Doubt is important for freedom. It distinguishes a democracy from a monolithic society. It distinguishes mice from men. This method of doubt is shown in the following passages:

J. Bronowski, in *Magic, Science, and Civilization* (1978):

> The scientific revolution (1500-1700) is comparable to the agricultural revolution. It gave a new way of thinking and looking at the universe. The scientific way of thinking is a human way of thinking. The great human revolutions are intellectual revolutions.

Doctor E.O. Wilson, *On Human Nature* (1978), has been interested in the biological basis of human behavior. He suggests that there has been an evolution of behavior. He asks if altruism, morality, religion, and love are survival strategies of our genes, biologically evolved over millions of years before we were human. He is asking questions that will bedevil us in the twenty-first century. That there is an evolutionary basis for human behavior remains to be established. Notice, however, Wilson's view of the spirit of science.

> Opinion on the merit of these arguments will no doubt be as sharply divided as it was on the sections dealing with human behavior in *Sociobiology.* At the risk of surrendering advantage to those whose beliefs leave them no option but rejection, I wish to say the following to others who are prone to read this book uncritically as a tested product of science: I might easily be wrong—in any particular conclusion, in the grander hopes for the role of the natural sciences, and in the trust gambled on scientific materialism. This qualification does not represent false modesty but instead is an attempt to maintain strength. The uncompromising application of evolutionary theory to all aspects of human existence will come to nothing if the scientific spirit itself falters, if ideas are not constructed so as to be submitted to objective testing and hence made mortal. The social sciences are still too young and weak, and evolutionary theory itself still too imperfect, for the propositions reviewed here to be carved in stone. It is my conviction nonetheless that the existing evidence favors them and through them the broader confidence in biological inquiry that forms the main thrust of this exposition.

As an undergraduate, I took a course in comparative religions and used as one of my texts, *The Religions of the World,* by G.A. Barton (1930). This was one of the books published by the University of Chicago Press in its publications on religious education. Permit me to quote the editors. I have underlined certain phrases that were significant to me.

> This series of Handbooks is intended to set forth in a readable form the results of <u>the scientific study of religion and ethics</u>. The various authors do not undertake to embody in any detail the processes which lie back of their conclusions. Such <u>technical treatment is more appropriate for works of a strictly scientific character</u> than for those intended not only to be used as textbook and collateral reading in colleges and theological seminaries, but also to be of help to general readers. The volumes all seek to conserve the values of past religious experience. While each author is free to present his own conclusions, the entire series has the common characteristic of historical method. <u>The editors have not prescribed any rigorous uniformity of treatment, but believe that the individuality of treatment will serve to stimulate thought and discussion. It is hoped that the series will help to show that the method of experiment and criticism contributes to stronger religious faith and moral idealism.</u>—The Editors

The Acceptance of Scientific Ideas

J.B. Gurdon published a paper called "Transplanted Nuclei and Cell Differentiation" in *Scientific American* (1968). In this paper he described experiments in which he removed the nucleus from an intestinal cell of a tadpole and transplanted this nucleus into a frog's egg whose own nucleus had been removed. The egg with the new nucleus underwent development and became a frog that resembled the "parent" from whom the transplanted nucleus had been obtained. This technique has been called "cloning." By such a technique it is possible to obtain many frogs of the same genetic constitution (identical frogs). The methods by which Gurdon did these experiments were described in scientific journals. Other scientific laboratories have repeated and corroborated these experiments. Because Gurdon's work has been corroborated and because it can be explained by present-day concepts in genetics and embryology, it is accepted as "fact" by the thousands of biologists who have not done the experiment.

This experiment by Gurdon has scientific implications for biology, because it gives us a technique to probe the mechanisms of development. It also has considerable humanistic implications, because if new frogs can be obtained in this fashion then other species could also produce such progeny. If a man wished to have a child just like himself, then he could have a nucleus from one of his cells implanted into an enucleated egg and thus produce a son.

David Rorvik, a respected scientific journalist, wrote a book, *In His Image, The Cloning of Man* (1978). He tells the story of a wealthy and influential man who wished to have a son that looked like him. He therefore made arrangements through Rorvik to have the nucleus of one of his skin cells implanted into an egg from a female to produce a son. Rorvik claims that the events described in his book actually happened and that a child, two years old at the time, actually exists. The biologists at that time thought that while eventually cloning in man could be achieved, it was not yet possible. They considered Rorvik's book to be fiction. Rorvik said that he was sworn to secrecy and therefore could reveal no more information.

Should we accept Rorvik's book as "scientific fact"? Should we conclude that Rorvik is a fraud? How do scientists make up their minds concerning the validity of reported observations?

Compare the acceptance of the work of Gurdon and the skepticism regarding the work of Rorvik.

A SUGGESTION. Every scientist has had the experience of collecting unexpected data. You may find it interesting to inquire of your professors concerning startling data and why these data were seemingly bizarre to them. As a teacher you may find it interesting to collect these "scientific anecdotes." For my own part, I still do not understand how the earth can be a sphere and why I don't fall off.

CHAPTER 3

THE CONCEPT OF EVOLUTION, A BIG IDEA

I HAVE FOUND it informative to ask citizens "What is evolution?" One answer is, "Evolution states that man came from monkeys over thousands of years, but the Bible states that man was created by God in one day." This is a good and meaningful definition of evolution for many people of strongly religious inclination, because it gets to the heart of a central concern, "What is the relationship of man to his Creator?" To many people, evolution is a competing religion. It is, however, unfortunate that the concept of evolution, which was proposed and developed as a scientific concept and by scientific methods, should have been made into a religious matter. Given the religious implications of the concept, however, it was inevitable that this would occur. It is not unusual to be asked, "Do you believe in evolution, or do you believe in God?" One needs to ask, "How does the biologist view the concept of evolution?" To the biologist, the theory of evolution is a scientific concept. It attempts to make a true or false statement rather than good or evil implications.

Is evolution a fact, or a speculation? To the biologist, evolution is a process. Evolution is a characteristic of living things, like the process of growth, or respiration, or cell division. It can be studied by observation and experimentation, just like other biological processes. Evolution is the process by means of which infinite variety among species and organisms has occurred and is occurring.

"Do I *believe* in evolution?" A belief may be based on evidence or on faith. To me, evolution is not a matter of faith. If scientific data were marshalled to show that evolution is not true, then I would give up the concept. It is not my purpose to get you to "believe in evolution." I am not in the proselytizing business. It is my responsibility to inform you of the observations that established evolution as a biological process. Whether you accept these observations as evidence is a personal matter. I am never asked whether I believe in respiration or cell division. I think that based on scientific evidence that has been accumulated by thousands of scientists over the past 120 years that the concept of evolution is true. Stephen J. Gould, a paleontologist, states the current opinion of biologists: "The concept of evolution is as well established as the concept that the earth rotates around the sun."

You may find it interesting to inquire into the meaning of the word "theory." Theory may mean "speculation" or it may refer to a "concept that encompasses a body of knowledge." The theory of gravity is not meant to be a speculation as to whether or not gravitional forces exist but, rather, it is a description of gravitational forces and an explanation of gravity as it relates to the earth, the moon, and the rotation of the planets. What do you think that the "theory of medicine" means? What does the "practice of medicine" mean?

The universe is not static, it is constantly changing. And each new day the world is different than it has ever been before. There are various kinds of evolution that are interrelated. Cosmic evolution deals with the origin and changes in the universe, of galaxies, of suns, of planets. It is studied by astronomers and astrophysicists. Chemical evolution involves the origin of atoms from primitive particles and molecules from atoms. It is studied by chemists. Geological evolution studies the origin of the earth, the rising of mountains out of the sea, the destruction of mountains by erosion, continental drift. It is studied by geologists and oceanographers. Biological evolution deals with the origin of organisms and the changes in species that occur with time. All of these kinds of evolution involve changes that have occurred over vast periods of time, millions of years, and by natural processes.

Biological Definitions

Ayala, an evolutionary biologist, defines evolution as follows: "The millions of diverse living species we find around us in the modern world are all descended from a common ancestor that lived in the remote past. The processes that have brought this diversity about are collectively called evolution." The concept of evolution starts with the observation that there is an enormous diversity of species living today. It attempts to answer the question, "Where did these species come from and how did they arise?" The concept of evolution is an important principle in science and therefore appears in textbooks and is taught in biology classes. Let me present some further definitions.

1. Present-day organisms arose from ancestors that were unlike them. This is Charles Darwin's idea of "descent with modification." The ancestors of present-day mammals were different kinds of mammals which, in turn, arose from ancestors that were reptiles. Man arose from mammals that were not men.

2. All organisms living today have common ancestry. The starling, the sparrow, and the black snake are related, but the sparrow and the starling are more closely related, having a common ancestor in the more recent past, while the sparrow and black snake are less closely related, having a common ancestor in the more distant past. In a family tree, brothers and sisters are more closely related than first cousins, since they share a more recent common ancestry. A phylogenetic tree also shows these genetic relationships. The present-day classification of organisms is based upon phylogeny. We assume that those species that look more closely alike have ancestors in the more recent past. When we construct a natural classification, we try to arrange species with respect to the closeness of their hereditary relationships.

3. Species are dynamic units. Changes occur in populations with time and space in both the appearance (phenotype) and genetic construction (genotype). Evolution involves changes in the genetic composition of populations with the passage of generations. Biologists have observed these changes in natural populations (such as the peppered moth), caused these changes by use of mutagens, and directed these changes by controlling the environment. Biologists have not only observed changes within a species, they have made new species in their laboratories.

The study of evolution involves its consideration as an historic process. This requires (1) the reconstruction of events in the past; events that have never been observed by man. Such studies deal with layers of rocks, the movement of mountains and oceans, fossils, radiometric dating of rocks and fossils (geology and paleontology), and (2) the distribution of organisms in time and space (biogeography). We are concerned here with the grand sweep of evolution over millions of years. We attempt to ask questions as an historian or archeologist who looks for lost cities and asks, "Was there ever a Babylonian Empire; when did it exist; what was it like; what was its cultural impact on Western civilization; why did Babylonia become extinct?" Historians and archeologists search for tablets and artifacts and go on digs. Their conclusions are based on the available evidence. The study of evolution deals not only with history but also with the mechanisms of change. This involves the study of inherited variation (genetics, molecular biology) and the interaction between species and their environments (ecology). In these studies much use is made of statistics and mathematics. The fact of evolution as an historic process is accepted by the biological community. The concept has been thoroughly tested and is now considered an important principle. The mechanisms that cause evolution remain to be established and are under intensive investigation. That we do not as yet fully understand the mechanisms that cause evolution is independent of the observation that evolution has occurred. We can observe that a chicken develops in and hatches from an egg, but we do not completely understand the forces that cause this to happen. It is because scientists do not fully understand evolutionary mechanisms that they often disagree with one another and that the field of evolution is so exciting a research area today. Human gross anatomy is a precise descriptive science, but there is little research, because almost everything is known.

Special Creation

The antithesis of biological evolution is special creation. There is a Creation Research Society that is based in California that studies and teaches "scientific creationism." Its teachings are as follows:

1. Each "kind" was created separately. These kinds (= species?) separately appeared, complete, organized and adapted to their environment. God made a man, Adam, from the dust. "Kind" is not equated with species to the creationist, and, therefore, observable changes within species or from one species to another is not evolution. These changes represent variations within a kind. It is not clear, therefore, whether the unit of classification comparable to a kind is a species, genus, family, order, class, or phylum. Do all of the known species of fishes belong to the same kind? Do sharks belong to the same kind as salmon? Are horses and cows the same kind? Are donkeys and zebras the same kind?

To state that each kind was created separately can in itself be either a scientific statement or a religious statement. The question is, "On what evidence do you base your conclusion?" Is it based on observation and experimentation? Is it interpreted by natural laws? Scientific creationism is a religion. The fact that it is a religion does not mean that it is not true. It does mean, however, that it is not science and cannot be subjected to scientific investigation. Indeed, it is useless for scientific investigation. The creationist view is that each kind was created by a creator by means that transcend natural law and such creation can never be explained by science. Creationists state that creationism is a religion, but that Evolution is also a religion and therefore students should be presented with both of these belief systems so that they can make a choice. The creationist view is based on a literal interpretation of the King James I version of the Bible.

2. The earth is relatively young in age, about 6,000 years old. This age is based upon counting back through the generations (the begats) in the Bible. The original statement by Bishop Ussher of Ireland (1654) was that the earth was formed in 4004 B.C. and that anyone who denies this commits heresy. There is so much evidence and agreement among scientists in such diverse fields as geology, geochemistry, astronomy and oceanography that the earth is very old (billions of years) that this "young earth" idea may be difficult to support. Among creationists there is a group of people who think that the old earth idea may be more accurate. The mainstream of creationism, however, supports a young earth. This is an important issue among creationists, since the concept of an old earth is contrary to the view that the Bible must be read in a literal sense and that the Bible is a book of history.

It follows from this that "kinds" cannot be genetically related. They do not have common ancestors. The way that biologists classify organisms does not reflect ancestral relationships but is merely a logical series. It also follows that kinds cannot change into other kinds. Each kind is a fixed unit.

It needs to be pointed out that biologists of 100-200 years ago viewed a "kind" as a species. The great classifier Linnaeus thought that species do not change; each species is a fixed unit. He combined this with a religious view: "There are as many *species* as were created in the beginning by the Infinite." The special creation view was held by almost all biologists (and all people) before the time of Darwin. At one time even Darwin held that point of view, and he must have gone through a great amount of mental anguish to give it up. The present-day creationists object to evolution, because it seems to contradict the literal interpretation of the Bible. At one time no one questioned the literal interpretation of the Bible. The Bible was historic fact and not myth. If you believe in the literal interpretation of the Bible, then you cannot accept the idea of evolution. Because the Bible is a historic record, at one time there was a worldwide flood that involved Noah and the Ark.

Henry Morris in *The Remarkable Birth of the Planet Earth* points out the religious implications of evolution: (1) the Bible clearly teaches that all things were created in six natural days several thousand years ago; (2) belief in evolution is a necessary component of atheism; (3) belief in special creation has a salutary influence on mankind, since it encourages responsible obedience to the creator; (4) belief in evolution leads usually and logically to rejection of the trustworthiness of the Bible and, therefore, failure to appropriate its promises leading to salvation and eternal life. To Morris, evolution is clearly a moral and religious matter. Doctor Morris and colleagues have criticized the scientific data that favors the concept of evolution. This is certainly a legitimate thing to do. They have had no impact to date on the scientific community.

The evolution-creation controversy is not a scientific controversy, since evolution is an attempt to explain the origin of species and the infinite variety of species by natural laws involving the inherent properties of populations of organisms, while creationism attempts to explain the origins of "kinds" by divine intervention. For over 120 years, thousands of scientists have tested the concept of evolution. This has involved a critical analysis of the evidence not only by biologists but also by geologists, chemists and by independent methods. If one wishes to look upon the Bible as a scientific text rather than as an inspirational statement of the relation of man to God, then one must open the Bible to the same criticism and debate that one applies to scientific ideas. One must ask concerning apparent biblical inconsistencies and concerning the origin of the Bible. One must inquire into similar creation stories that preceded the Bible. And one must be permitted to carry on such inquiry free from threats of heresy. The evolution-creationism controversy is not a bona fide scientific controversy. The issue is one of biblical literalism. Is one to accept the English translation of the Bible as being word-for-word true, or is one permitted to interpret these stories in ways that are meaningful to each new generation? The question of biblical literalism is not restricted to the Judeo-Christian tradition but is found also among Orthodox Muslims and Hindus. The creation-evolution controversy is a two-edged sword. It not only challenges evolution but also biblical accuracy.

There are other speculations concerning the origin of life and man. One of these is that life has always existed on earth; another is that life came to earth from outer space. Christianity and Judaism represent only two out of many religions that have a creation story. During the past 400 years, science and Scripture have come into conflict on many different scientific ideas. The idea of evolution is simply another instance of this.

The idea of evolution has appeared at various times before Darwin. The best known proponent was the famous French biologist Jean Baptiste de Lamarck (1749-1829). Lamarck proposed that the resemblance of organisms arises through evolutionary modification. Animals that resemble each other more closely have closer ancestors. The credit for the idea of evolution, however, is usually given to Charles Darwin. Before Darwin, most people were unaware of the possibility of evolution. What Darwin did was to not only present the idea but also to marshall an enormous amount of evidence in support of the concept of evolution AND to propose a mechanism to explain how evolution could occur. In addition, the time and conditions were ready for it to be accepted. Up until the nineteenth century, scientists in many European countries were afraid to present ideas that were contrary to Scripture, and people who proposed such ideas were intimidated and forced to recant their views. Indeed, in order to be granted a degree from Oxford University, it was required to pledge that one accepted the Bible as an accurate view of history. In England, at the time of Darwin some essays had been published by well-respected Anglican clergy warning that certain scientific ideas that were developing relating to the age of the earth were contrary to Scripture, and that clergy would need to modify scriptural interpretation to encompass these scientific ideas. In addition, there developed in Germany and in England a reassessment of the literal interpretation of the Bible. Biblical criticism developed as an important research area. It was pointed out by certain more liberal clergy that the strict literal interpretation of the Bible made no sense (Ruse, 1979).

When Darwin's *On the Origin of Species* appeared in 1859, it entered a Victorian climate in which there already was a full-scale battle concerning biblical literalism. One should pause for a moment to reflect upon the fate of Darwin had he published his book in sixteenth century Italy. One should pause for another moment to reflect that science is a social and humanistic activity. The acceptance of scientific ideas into a society depends to a large extent upon the current beliefs and prejudices of that society.

A major problem in accepting the idea of evolution was the belief that the earth was only 6,000 years old. Even if species do change (as Darwin had proposed), they change so imperceptibly that 6,000 years is not enough time for new species to arise. The age of the earth needed to be determined before biological evolution could be considered. James Hutton, in 1795, published *Theory of the Earth* and concluded that changes in the earth's surface were continuously taking place by natural causes (water, wind) over a period much longer than 6,000 years. The earth is really quite old. Charles Lyell (1830), in his *Principles of Geology,* also recognized that the earth was millions of years old. Lyell's book gave Darwin "the gift of time," the many years that were needed for evolution to take place. Biological evolution starts with geology. The idea of evolution is incompatible with "a young earth."

Darwin had been working on his idea of evolution from 1833 to 1858 when he received a paper written by Alfred Russell Wallace, a naturalist in the East Indies. Wallace arrived at the same conclusions as Darwin had from his studies on the distribution of animals (zoogeography). Darwin was surprised and dismayed to find that he was "scooped." Darwin consulted with friends who knew about his efforts for 25 years, and they proposed that papers by both Wallace and Darwin be read at the same meeting of the Linnaean Society of London. On July 1, 1858 the secretary of the society read the two papers submitted to the society proposing the theory of evolution of species by means of natural selection. These papers apparently caused no unusual excitement. Professor Thomas Bell, then president of the society, expressed regret in his 1858 annual report that 1858 had "not been marked by any of those striking discoveries which at once revolutionize, so to speak, the department of science on which they bear." It was a classic example of "putting one's foot into posterity" (*New York Times,* July 2, 1958). New ideas are not immediately recognized! It takes time for them to be assimilated!

Did Darwin discover evolution? Who should get credit for a scientific discovery? In the study of the natural world, different people sometimes independently arrive at the same idea. In science, one generation builds upon another. In the scientific community there is an exchange of ideas among people working on the same problem. Sir Isaac Newton, the discoverer of gravity, recognized this. "If I have seen farther than others it is because I have stood on the shoulders of giants." If Darwin and Wallace had not discovered the concepts of evolution, some other biologists would have made this discovery. Laws of nature exist. They are "out there" waiting to be discovered by the receptive mind and by a society that tolerates new ideas.

Implications for Science

The concept of evolution is the most important unifying principle in biology. It molded biology into a single science, because it gave to botanists, zoologists, and bacteriologists a major hypothesis that needed testing. It is a unifying principle, because it relates to all aspects of biology. Evolution explained many observations that otherwise made no sense or otherwise seemed to be unrelated. These included the similarity of bone structure among frogs, birds, whales and men; the presence of gill slits in mammalian embryos; the existence of a useless pelvic girdle in whales. The idea of evolution gave meaning to the sequence of fossils found in layers of rock. It changed the study of the classification of organisms from a cataloguing job into an attempt to classify in such a way as to show genetic relationships. With the development of genetics and molecular biology, studies are being pursued relating to similarity of genes and of proteins in different groups of organisms. The early biologist was concerned with homology (similarity) of structure among organisms based on their similarity in development. The molecular biologist studies the chemical homology of proteins based upon the similarity in the genes (DNA) and determines the degree to which similarity of proteins among species can be correlated with their phylogenetic relationships. Biochemical evolution is a major field of study. There are newly established scientific journals that deal with biochemical and molecular evolution.

The concept of evolution is currently of importance to psychologists. The discrete emotions of human experience are very much a part of their biological heritage. Darwin has been quoted by psychologists more in the last 10 years (1975-1985) than in all of the previous 100 years. The study of fossils continues at a rapid rate by paleontologists; work on primitive man involving both biological and cultural aspects is an active field in anthropology. Changes in mountain ranges, in oceans, the movements of large land masses represent exciting and observable phenomena. Astronomers study a universe that has been changing over billions of years.

Implications for Western Society

The concept of evolution is a major contribution to human thought not only because it changes how we describe and study the universe but also how we view man and his relation to the world around them. It gave a materialistic explanation for the origin of species in contrast to a supernatural view of creation. It proposed that man arose from a different species of animal. It showed that man is a part of nature. Whatever his "human" attributes he is also an animal and is related by common ancestry to other animals. Not only are all men brothers, but all animals are brothers. Evolution revealed an apparent conflict with Genesis. This caused, and is causing, spiritual unrest. It forced Western people into a real identity crisis. It made them consider again "what is their relation to a God" and "what are people's purpose on earth." Evolution raised the question concerning how to get at the truth, by belief or reason, by spiritual experiences or observation and experimentation, by supernatural or by natural explanation. Evolution was the final blow to free science from religious (scriptural) authority. Just imagine, if whenever you learned about a new scientific idea that you had to clear it with your church. "Reverend, is it all right with you if I believe that man is genetically related to chimpanzees?" Or, if you were a scientist and made some new observation, would you say, "I have these data and draw these conclusions from them. I had better clear this with religious authorities to find out if I can accept them." Evolution increased man's world view. He could stand up and look at the universe around him. He could question; he could challenge; he could innovate; he could think. It made man grow up and look to himself to solve his problems. With such awesome implications, no wonder Darwin waited until 1859 to publish an idea that he had conceived in 1835. Would you have had the guts to publish?

"If we choose to let conjecture run wild, then animals, our fellow brethren in pain, disease, suffering, and famine—our slaves in the most laborious works, our companions in our amusements—they may partake of our origin in one common ancestor—we may all be melted together." (This view was written by Darwin when he was 28 years old; from Loren Eiseley's *Darwin's Century.*)

The Relation of Evolution to Other Fields of Knowledge

Let us inquire into the impact of the concept of evolution on human thought. What is the breadth of its influence? If you are a university student, inquire concerning the influence of evolutionary theory on the disciplines of geology, anthropology, psychology, sociology, biochemistry, physics, history, agriculture, medicine, religion, literature, astronomy, geography. You need to recognize that evolution is not simply a biological matter. How many scientific journals deal with evolution? List these journals. If you read the newspapers or certain magazines, you will find articles dealing with evolution. Look at the book review sections of newspapers. If you are startled by the breadth of evolutionary theory, you have been leading a very sheltered life.

What Do We Mean By "A Theory"

The cell theory is one of the major concepts in biology. It states that all organisms are made up of basic units, called cells. Is the cell theory a speculation? At one time it *was* a speculation. Two biologists, Schleiden and Schwann, proposed this theory (as speculation) in 1838. It was based upon their observations and the observations of other biologists, since 1665, that all of the plants and animals examined were made up of cells. This speculation (a testable hypothesis) stimulated other biologists to examine a wide variety of species. In every instance, these organisms were found to be composed of cells. As research continued in many laboratories, it was found that all cells carry out the same chemical and physiological functions of life; that all cells are basically the same in structure, consisting of a nucleus containing chromosomes and a cytoplasm containing mitochondria and an endoplasmic reticulum; that all cells arise by means of mitosis from preexisting cells. It was found that an egg is a cell, a sperm is a cell, and that these gametes fused to form a zygote, which is a cell.

Today, every biologist accepts the cell theory as a valid generalization. The term "theory" is used to describe *a body of knowledge* relating to cells rather than *a speculation*. The cell theory has led to the development of the fields of cytology, histology, tissue culture. It has been important in the understanding of genetics, embryology, physiology, tissue culture, evolution, cancer. Yet, there are limitations to the cell theory. Bacteria do not have a nucleus separated from the cytoplasm by a membrane, viruses do not possess a nucleus, mature red blood cells of mammals have no nuclei, ameba have been considered non-cellular organisms. Is the slime mold a cellular organism? Can we say that all organisms are made up of cells when only a relatively few out of millions of species have been examined? We deduce from the cell theory that if we were to examine a newly discovered species of fish, that it would be made up of cells. It would be exciting to discover a species of fish that was not made up of cells. This would not destroy the usefulness of the cell theory, but it would open up a new area of research.

The Teaching of Evolution and Special Creation

Many attempts have been made to have the concept of special creation taught in public schools alongside the concept of evolution. How shall this lesson be taught? Shall we teach it in science courses as two conflicting and alternate hypotheses? If so, one needs to marshall the scientific evidence in support of the theory of evolution and the theory of special creation. We will need to consider as theory that there was once a man, Adam, who lived and sinned in the Garden of Eden. In a scientific classroom special creation theory is no holier than evolutionary theory. If you wish special creation to be taught as *special creation,* then teach it in Sunday School.

There is considerable objection to the teaching of evolution as "absolute fact" instead of as theory. Why do biologists hold on to the theory of evolution after 125 years of intensive research? Is it only absolute facts that should be taught in the classroom? What is the process by means of which a theory becomes a fact? Are there any absolute facts in science? Are there absolute facts in religion? Should there be objections to teaching as fact that genes consist of DNA or that respiration involves the utilization of oxygen? Who is to decide that one theory is fact and another is speculation?

If one is to teach evolution and special creation in the classroom, one must clearly separate the scientific arguments from the religious arguments (evolution is false because it is immoral; if evolution is true then there is no meaning or purpose in life; if man was not created, then there is no such thing as sin; evolution is a hoax put forth by atheists).

If evolution and special creation were taught in the classroom, it should have as a major goal to consider the differences between a scientific and a religious approach to truth. We would need to consider the nature of scientific evidence and the nature of religious evidence. We would need to consider the role of the scientist and of the clergyman in society. Why should we believe scientists? Why should we believe clergymen?

William Jennings Bryan, a presidential candidate in 1896, 1900, and 1908, believed in the literal interpretation of the Bible. Bryan believed that the theory of evolution put forth by Darwin denied the biblical story of man's divine creation and, therefore, should not be taught. Should evolution be taught in the science classroom?

It would be stimulating to learn in this way. Such a classroom experience, however, requires a degree of tolerance not now visible in our society. It would be a challenge to the art of teaching, to the educational process, and to the open-mindedness of the community.

Some Questions to Consider

Evolution should not be taught in science classes in the public schools because:

1. it is contrary to Scripture
2. it destroys faith in the Bible
3. it is an immoral concept
4. it is only speculation
5. it is not true
6. it is dangerous knowledge for children
7. it will influence people to behave as animals
8. it will increase violence, crime, and sin in society
9. it will destroy all hopes of salvation
10. it is offensive to many people

Evolution should be taught in science classes in the public schools because:

1. it is an important scientific principle
2. it is a concept that has been tested and developed by the scientific community
3. it is a characteristic of living things
4. it is an important scientific speculation

Special creation should be taught in science classes in the public schools because:

1. it is a true explanation of the origin of species
2. it is a scientific explanation of the origin of species
3. it bolsters man's image of himself as a divine creature
4. it gives counterbalance to the concept of evolution
5. it is a matter of academic freedom

Special creation should not be taught in science classes in the public schools because:

1. it is not science
2. it is not true
3. it is contrary to scientific observations
4. it is demeaning to the Bible

Both evolution and special creation should be taught side by side in (1) science classes, (2) history classes, and (3) Sunday school classes.

The Garden of Eden Story should be discussed in (1) science classes and (2) Sunday school but only as a theory. The Garden of Eden should be presented along with the hundreds of other creation stories that have been proposed by different civilizations.

Only concepts that have been established as fact should be taught in (1) science classes and (2) Sunday school.

A teacher who is a creationist should not be required to teach evolution in science classes, because this concept is contrary to personal conviction. It is a matter of conscience.

How does the science teacher determine whether a scientific concept is a speculation or a fact?

DO IT YOURSELF. Compare and contrast, by means of a chart, the concepts of evolution and creationism.

CHAPTER 4

NATURAL SELECTION, A PROCESS

CHARLES DARWIN had the idea of evolution as an historic process, but he initially had no explanation to account for the mechanisms by means of which evolution takes place. He recognized that change occurs with time in successive generations of organisms and that populations of organisms exhibit certain features that enable them to exploit their environment. Whales, although mammals, have a fish-like shape and limbs that can be used as paddles. He believed that gradual changes in structure and function occurred to give rise to populations with adaptations for particular environments. He recognized that heredity forces were involved in adaptation. The biologist Dobzhansky stated that, "Next to understanding life itself, adaptation is the most difficult problem."

Darwin finally arrived at the concept of natural selection to explain the process of evolution and adaptation. He was influenced by the domestic breeders who could make new varieties by artificial selection. These breeders developed the modern breeds of cattle and pigs by breeding their animals to the desired types (line breeding). Over a relatively short period of time there were developed 140 varieties of pigeon; 35 varieties of pigs from the wild boar; 125 varieties of domestic fowl; over 300 varieties of dogs from the Mexican Hairless to the Boxer, to the Dachshund, to the Collie, to the Great Dane; varieties of brussel sprouts, cauliflower, broccoli from the wild cabbage. Darwin recognized that these domesticated varieties had common ancestry and that there was a great inherited diversity within each species. If we did not know the breeding history of Dachshunds, Chihuahuas, Great Danes, and Airedales we might place them into different species or even genera. He reasoned that what artificial selection (selection by man) could accomplish over hundreds of years, natural selection could do over longer time periods.

Darwin was influenced also by *Essays on Population* written by the economist Malthus. It is interesting that Malthus had also been read by Alfred Russell Wallace, co-discoverer of evolution. Malthus stated that, "Human populations tend to increase in numbers in geometric progression." "More children are born than live to maturity, because of the struggle for existence; hunger, war, disease keep the population numbers in check." Malthus was pessimistic concerning what could be done, because (1) "Food is necessary for the existence of man," and (2) "The passion between the sexes is necessary and will remain nearly in its present state." "These two laws ever since we have had any knowledge of Mankind, appear to have been fixed laws of our nature, and as we have not hitherto seen any alteration of them, we have no right to conclude they will ever cease to be what they now are" (from Hardin).

Charles Darwin's theory of natural selection can be stated as follows:

1. Organisms have the potential to overproduce. A single oyster or salmon produces millions of eggs. If all of these eggs were fertilized and developed into mature individuals, the world would within a single generation be inundated with oysters and salmon. Even for slowly breeding species such as man, there could be an overpopulation because of the potential for geometric increase from one generation to the next.

2. Despite this reproductive potential, the population size for each species remains fairly constant; because most of the gametes are not involved in fertilization, most of the fertilized eggs do not reach reproductive maturity. Each new generation arises from only a relatively small number of potential parents.

3. Within each population all organisms are not alike. There are variants. Darwin recognized that different types existed in a population, but he did not know how this variation was started or how it was passed on to future generations. At the time of Darwin, the field of genetics had not yet been developed.

4. Those variants that were better-suited for a particular environment survive and pass their characteristics on to progeny. Those variants that are acted upon unfavorably by the environment either perish or produce less progeny. Thus, over a period of generations there occurs a change in the characteristics of a population and a general improvement of the population in a particular environment.

In the scheme of natural selection, there is a selection for those variants that are better adapted. For changes in populations to occur, what is required is genetic variation within a population and environmental change.

The experimental proof for natural selection had to await the development of the fields of genetics and ecology. Scientists have found that selection mechanisms as proposed by Darwin exist in nature and are involved in adaptation. Natural selection, although important in the evolutionary process, may not be the only mechanism involved. This area involving adaptation and speciation is today an active field of inquiry.

Impact of "Natural Selection" on Society

The theory of natural selection has had a great impact on the social sciences. There developed a large amount of work on "Social Darwinism" in which the rule of natural selection was applied to the rules and conduct of society. The concept of natural selection brought forth images of an active struggle among organisms within the same species for survival ("survival of the fittest" and "red in tooth and claw"). This idea disturbed the Victorian society who considered it to be too harsh to be a law of God and nature. It was realized that competition may involve not only an active struggle between organisms but also a passive struggle between an organism and its physical environment (penguins have insulation against the cold, desert plants have spiny leaves to conserve water). It was recognized also that evolutionary success involves reproductive success. The strongest may not produce the most offspring.

On the other hand, the idea of natural selection was used to justify colonialism, wars, exploitation by ruling classes, slums and poverty. It was used by Andrew Carnegie to exploit labor in the steel mills. The plea of immorality did not stop the exploitation. "We are just following the law of nature." A society was viewed alternately as a competitive and a cooperative unit. Can one select for variants involving cooperation in a group? Are cooperation and altruism under genetic control and explainable by natural selection? The field of sociobiology is presently developing as an outgrowth of Darwinism. It deals with the evolutionary basis of individual and group behavior. Its impact will be determined in the twenty-first century.

Is There a Purpose in Evolution?

Is the means by which evolution occurs directed towards a specific goal? Evolution could be more acceptable to many people provided it had a long-range purpose and ended with man as the highest type of organism. One who wishes to find a teleological explanation looks for a mechanism that gives a long-range purpose to evolution. Natural selection does not appear to be such a mechanism. It is concerned with a response to the immediate environment. It cannot guess what will be the environmental conditions of the future. In the scheme proposed by Darwin there appears to be the element of chance. Can chance be a determiner of evolution? Even Darwin pondered whether a complex structure such as an eye could have developed by chance. If the whole business were to start over again, would we once again end up with man? Where is the orderliness in the universe under this scheme? To many people, the orderliness of the universe and the complex adaptations show that there is a design. The argument from design that is used by theologians and philosophers states that if there is a design, there must be a designer. This design in nature relates to a God that constructs and directs the order of the universe. Natural Selection cannot be a design, because it lacks long-range purpose. One can answer that "chance" is not the same as "accident." Chance is a design; it has order; it has predictability; it is indeed the basis for the transmission of genes. The field of Mendelian heredity is based on chance. Although one frequently cannot predict the heredity traits for a single individual, one can predict with accuracy the proportion of individuals within a population who will show certain traits. Chance is a demonstrable law of nature. Chance is a characteristic of living things. One can also argue whether God controls the universe directly or through his natural laws. God does not need to stand at the helm to direct the rotation of the planets. This can be done quite well through the forces of gravity. God does not decide which out of billions of acorns become oak trees, nor which out of billions of gametes become people. It is indeed the scientist and not the theologian who searches for design in nature and an orderliness in the universe. Many theologians prove the existence of a designer by the miracles that He performs.

To the scientist a design is a process. Natural selection is a process to explain evolution. It is an orderly process that enables populations to adapt to their immediate environment. This population adaptation occurs at the expense of many individuals. To the theologian, design refers to a moral purpose for which God created the universe. The universe was not created by chance but rather for some long-range goal. The word "purpose" is sometimes used in place of the word "function." What is the purpose (function) of respiration or photosynthesis?

Aldous Huxley in his poem *Fifth Philosopher's Song* addresses the element of chance in nature.

A million million spermatozoa
  All of them alive:
Out of their cataclysm but one poor Noah
  Dare hope to survive.

Out of all those possible people who could have been born, why me? Perhaps some other sperm would have produced a person who is worthier than I. My chances to be born were 100 million to one.

Robert Frost in his poem *Accidently on Purpose* is concerned with the purposefulness of the evolutionary process. Unlike Huxley who stresses randomness, Frost considers that the universe had a purpose from the beginning. Frost's poem places evolution in a teleological framework. If the whole business were to start over again, we would end up with people.

Never believe it. At the very worst
It must have had the purpose from the first

and

Grant me intention, purpose, and design—
That's near enough for me to the Divine.

Chance and purpose, design and randomness, individual order and statistical order, and determinism raise questions not only for scientists but also for poets, for philosophers, for theologians, and for citizens. The approach to the concept of chance by the scientist is different than the approach by the poet or the theologian. It is a crucial issue in religion, because it relates to whether the Creator is personally involved in each of us or whether people are just one great big fish tank to Him. Statistical order is emotionally unsatisfying. Religion has to mean a relationship that each of us has with God or it means nothing. It is for this reason that the theory of natural selection is disturbing to many people.

The role of poetry is to ennoble people and nature and to cry out against injustice. The role of science is to describe the laws of nature.

THE MUTATION-SELECTION CONCEPT

Charles Darwin proposed the hypothesis of natural selection to explain the process of evolution. In this hypothesis he assumed that there were variants in populations and that these variants could be inherited. He assumed also that some of these variants would be acted upon more favorably by the environment and would therefore increase in number, while less favorable variants would decrease. This would result in a new population that was better adapted to the environment. Biologists today recognize that inherited variation exists as mutations. They have developed a mutation-selection concept based on genetics, a field of knowledge that did not exist at the time of Darwin. Can one demonstrate this selection process in the laboratory, in nature, and in the affairs of man?

Replica Plating Test

Suppose we have culture dishes that contain 1,000 colonies of cells of the bacterium *E. coli.* Within each colony, all of the cells comprising it are identical, since they arose by division from a single initial cell. How can we determine whether some of these cells possess the mutant for resistance to a toxic substance, the antibiotic streptomycin, even though they have never been exposed to streptomycin or whether it is the antibiotic itself that changes sensitive cells into resistant cells. An elegant test for this problem was devised in 1952 by the geneticist, Joshua Lederberg. He reasoned that if all of the colonies are equally sensitive to streptomycin (there is no variation due to mutants) and some are changed to resistant colonies upon exposure to the antibiotic, then one would expect different colonies to show resistance. If, however, there is genetic variation, where certain colonies were already resistant to streptomycin before exposure to the antibiotic, then only the resistant colonies would survive exposure to streptomycin.

Lederberg devised a replica plating method whereby the bacterial colonies and their precise positions from the original plate could be copied onto other culture dishes. One would therefore now have many duplicate plates. The technique involved the pressing of the original plate onto a die covered by velveteen cloth. The little needles of velveteen picked up cells from each colony. Nutrient plates containing streptomycin but without *E. coli* could now be pressed onto the velveteen and thereby cells could be transferred to successive plates in exact positions. After growth for a day or two, duplicate plates were available for comparison.

It was found that the same colonies were resistant on all of the replicate plates, and the original plate. This proved to geneticists that these colonies already possessed the gene for resistance to streptomycin even though they had never been exposed to streptomycin. The inherited resistance had occurred by mutation from streptomycin-sensitive to

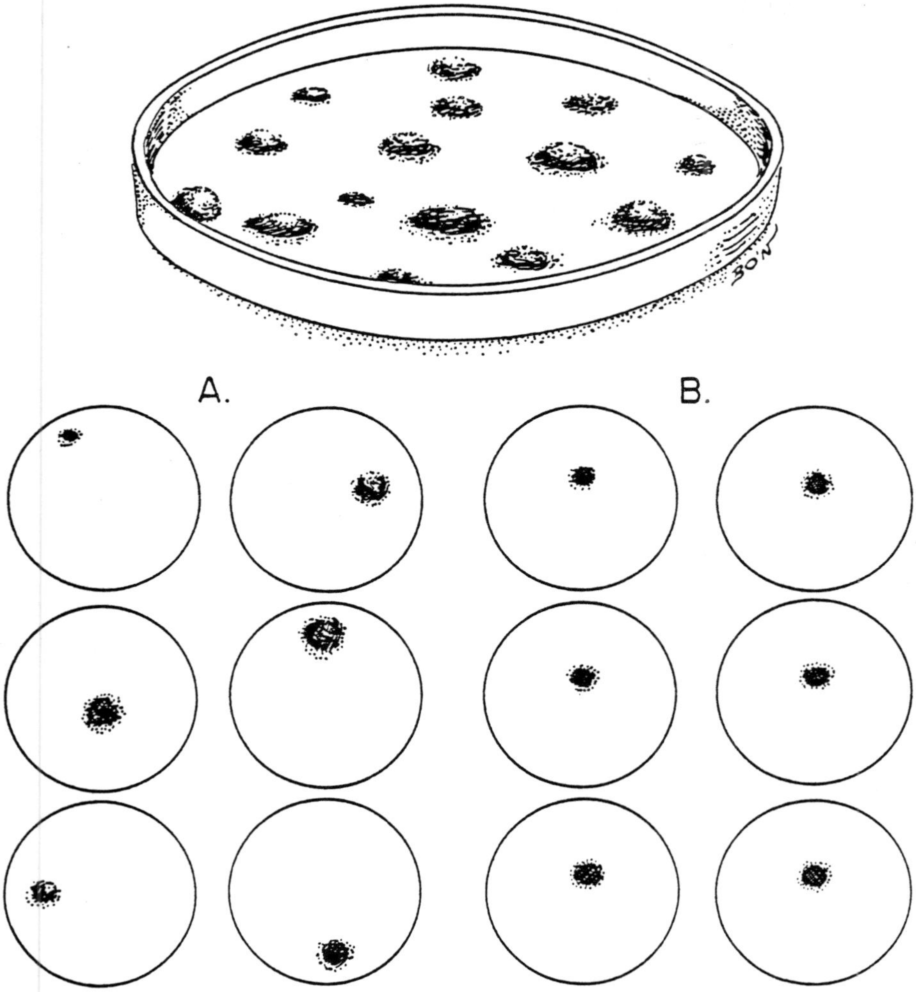

Figure 1. Replica plating as a test for the mutation-selection hypothesis. Here is a diagram of the original culture dish showing 15 colonies of *E. coli,* each colony occupying a definite position in the culture dish. By means of replica plating many copies were made. Only six cultures are shown here. If these cultures are exposed to streptomycin, we would expect that different colonies would survive in each of the culture dishes (result A) or the same colony would survive in each of the culture dishes (result B). It is result B that is obtained. Surviving colonies are shown by the solid circle.

streptomycin-resistant cells before exposure to streptomycin. The streptomycin was the selecting agent. What do you think of the conclusion reached by Lederberg? Is this a good test of the process of mutation selection? If you press your thumb against an ink pad, you can transfer your fingerprint many times to a pad of paper. You have done "replica plating."

Some Laboratory Selection Experiments

The bacterial cell *Escherichia coli* is present in the colon of man. This organism can be reared in a culture tube. If 100 million *E. coli* are placed into a culture tube and the antibiotic streptomycin is added, all but about 10 cells perish. The initial culture is said to be streptomycin-sensitive (S^S). The 10 cells that survive will multiply in the presence of the antibiotic and a new culture will arise that is streptomycin-resistant (S^R). If these organisms continue to reproduce in the presence of streptomycin, a culture is produced in which the *E. coli* survive better in the presence of streptomycin than in its absence. The cells have become streptomycin-dependent (S^D). Recognize that in this process in which the culture changes from $S^S \rightarrow S^R \rightarrow S^D$ that mutations have occurred, that selection for favorable mutants has occurred, that adaptation has occurred. New resistant cultures arise from the relatively small number of cells that already are resistant. In the presence of streptomycin, a substance that is toxic to the original culture, not only does a resistant culture arise but eventually one that incorporates the streptomycin into its metabolism.

Selection in Nature

There are many examples of selection in nature. The California orange trees are attacked by scale insects. Control of these pests has occurred through the use of smudge pots containing cyanide (HCN). Over the years, it has been found that the initial concentrations of HCN used for control of the scale insects were ineffective and that the HCN concentration needed to be raised.

Housefly populations that initially were almost eradicated by the insecticide, DDT, are now able to survive the initial concentrations of DDT that were used.

Different species that live together have an effect on changing each other's gene frequencies. This is shown in the case of rabbits that were brought to Australia. In 1859, twenty-four rabbits were imported from Europe. This population increased its range and its number so that by 1928 there were 500,000,000 rabbits. The rabbit became an agricultural pest by destroying pastures used for sheep grazing. The reduction in the rabbit population was brought about by inoculation with the myxoma virus coupled with mosquitos that served as vectors. Within three years, 95 percent of the rabbit population had been destroyed. But then, the rabbit population started to increase again. Genetic changes in both the virus and the rabbit population occurred. The rabbit populations became more resistant to the myxoma virus; the virus populations developed less virulent strains. By selection of appropriate genes the virus and rabbit could live together. There was a co-adaption. What advantage would a non-virulent virus have over a virulent virus?

In wheat and other grains, strains that are sensitive to (injured by) rusts and molds occur. The selection of rust and mold-resistant wheat strains in the grain fields enables it to survive, but this also causes selection for rusts and molds that can survive on the wheat. And so, a continuous shift in genes occurs as these different species interact with one another. As a standard technique, a wheat field containing many plants is sprayed with a plant pathogen. This results in the survival of wheat plants that are resistant to this plant pathogen. These plants are used to produce resistant populations. The agriculturist is using the concept of mutation and selection to obtain resistant strains.

Selection and Human Affairs

Most of the animals such as poultry and cattle that are reared for human consumption are fed an antibiotic. This results in the ability of these animals to grow at a faster rate. This addition of the antibiotics penicillin or tetracycline, however, may be potentially harmful to man because it allows for the selection against antibiotic-sensitive and the selection for antibiotic-resistant strains of pathogenic microorganisms. Thus, if an infection occurs in a person, he may not be able to combat it with further antibiotics. The Food and Drug Administration is now investigating whether the feeding of antibiotics to poultry and cattle should be permitted.

A serious public health problem exists in Africa and in Southeastern Asia, because millions of people are exposed to the malarial parasite *Plasmodium*. This protozoan parasite lives in the red blood cells of man and in the gut of the mosquito *Anopheles*. Plasmodium is transmitted from man to man by means of this blood-sucking mosquito. Control of malaria requires the elimination of the *Plasmodium* by means of the oral intake of the antimalarial drug chloroquine and the elimination of the mosquito by means of the insecticide DDT. These control measures have been of only temporary success. Malaria is now on the increase. The chloroquine-sensitive protozoans and the DDT-sensitive mosquitoes have been eliminated. The chloroquine-resistant and DDT-resistant organisms remain in the population and have increased in number. Since sensitivity and resistance is a gene-controlled trait, these genes are passed on to progeny that are now resistant.

Observations on resistance to antibiotics have important bearing upon public health. In the August 6, 1981 issue of *The New York Times*, there is a report on the need to control the use of antibiotics. An overuse of antibiotics such as penicillin and tetracycline has dramatically reduced their effectiveness. Bacteria repeatedly exposed to drugs become resistant to them. In the U.S., antibiotics are often indiscriminately prescribed. In underdeveloped countries antibiotics are available without a prescription and used as a cure-all. The reduced effectiveness costs lives. If you do not kill the bacterial population with the right dose of antibiotic the first time, you may not get a second try. If we were to routinely give everybody in the world penicillin, we would eventually have a population of people who harbor penicillin-resistant microorganisms. These organisms will have arisen by the selection scheme proposed by Charles Darwin. Suppose that someone discovered a "universal antibiotic," one that could destroy all species of microorganisms. What would be the consequences of this discovery? Should we recognize it is a boon to mankind, or should we organize societies to prevent its use?

Natural Selection as a Creative Process

It is often thought that evolution is the result of purely random processes and that natural selection is simply a sieve. Francisco Ayala makes the point that natural selection is a creative process, because it enables combinations of favorable genes to be brought together and thus it is an adaptive process. Mutation and recombination are random processes that serve to increase genetic diversity in populations. It is natural selection that gives direction and meaning to the evolutionary process. It is a way to build the best genetic combinations into populations.

The vertebrate eye, that remarkable organ, did not suddenly appear at random in its present state. Over a period of 1.5 billion years favorable genetic units have been added to the gene complex for making the eye and in a stepwise way. The ancestors of vertebrates possessed organs sensitive to light. This initial perception of light upon which a more sophisticated vision was built were more important for the survival and reproductive success of the species.

Natural selection is an ingenious mechanism for the survival and adaptation of populations. It preserves the best genetic combinations in a population in which the environment is stable. It exploits new genetic combinations in a population whose environment has changed. It promotes different gene combinations in diverse environments.

> Can it then be thought improbable, seeing that variations useful to man have undoubtedly occurred, that other variations useful in some way to each being in the great and complex battle of life, should sometimes occur in the course of thousands of generations? If such do occur, can we doubt (remembering that more individuals are born than can possibly survive) that individuals having any advantage, however slight, over others, would have the best chance of surviving and of procreating their kind? On the other hand, we may feel sure than any variation in the least degree injurious would be rigidly destroyed. This preservation of favorable variations and the rejection of injurious variations, I call Natural Selection. From Darwin in *The Origin of Species* (from Ayala).

Nancy Ann Dahl in her poem *Encounter with Evolution* is responding to the tragedy of the death of her firstborn son. Why did this happen to me? Her son carried a gene for achondroplasia, a kind of dwarfism that often makes breathing difficult. How does she justify the death of her son? "A random mutation each forty thousand." "If there were no mutations we would still swim in the sea."

> Prayer, omens, afterlife, gods.
> Flickering shadows from childhood, from history.
> Attempts to explain our random journey.
> Does need make reality or wishes make truth?
> No god commands evolution's walk.

This is her personal statement. Another poet may explain this tragedy as "God's will." Her poem raises the questions of individual worth. Does God have control over natural events such as mutation, fertilization, natural selection? Voltaire was concerned in *Candide* with trying to explain an earthquake in Spain that killed thousands of people and with the suffering of people in "This best of all possible worlds."

It would be comforting if every woman produced four eggs, every man produced four sperm and that four zygotes would arise, each developing into a healthy and viable boy or girl. But this is not what happens. There is a great amount of fetal death and postnatal death. Millions of people have died from the disease sickle-cell anemia to allow for the survival of their population. I suppose that we need to cry about and ennoble those things that we cannot control. We seem to cry but not enough about matters we can control such as the starvation of populations, infectious disease, illiteracy. Are these also "God's will?"

CHAPTER 5

SCIENCE AND RELIGION IN HISTORIC PERSPECTIVE

THE MAJOR objection to the concept of evolution comes not from science but from religion. By some people evolution is deemed false because it is contrary to Scripture. To evaluate this criticism one needs to look back into history and ask whether this sort of thing has happened before. Indeed, the history of scientific thought over the past 400 years shows that science has continuously tried to free itself from religious explanation and appeal to authority. This theme has been addressed in an excellent book, *The Death of Adam* by the historian, John C. Greene (1959). This book deals with evolution and its impact on Western thought and with the religious unrest that has been caused by the growth in the sciences. I will rely on this book for much of my discussion. During the 1600s, 1700s, and 1800s the concepts of change, space, time, and movement were developed. There was an emergence of astronomy, geology, and biology. Every one of these sciences had to struggle with harassment from religious authority. Science developed out of philosophy. There were moral philosophers concerned with the study of human conduct and there were natural philosophers concerned with the observation of God's world. Throughout this period people who worked in science justified their studies for the purest of reasons, to explain God's world and God's work, the heavens, the earth, the planets, the plants and animals. In most instances, the creation story and the Deluge (Flood) were believed to be historic events and the earth was considered to be about 6,000 years old. In astronomy, geology, and biology, all scientific interpretations had to be consistent with Scripture and those scientists, regardless of their avowal to Christian doctrine, who presented ideas that appeared to be contrary were reprimanded and often forced to recant. "Theology was queen of the sciences."

Since the universe had been made by God, it must be perfect and unchangeable. God would not make a mistake. Galileo, however, found craters on the moon and moving spots on the face of the sun and asked whether these were imperfections. Galileo found that heavenly bodies moved about and suggested that bodies in motion are as natural as bodies at rest. The objection to Galileo was that the world is already perfect and there was no need, therefore, for motion and change.

The most famous instance of science-Scripture antagonism was the trial of Galileo (1633). Galileo thought that the earth moved around the sun and this idea was considered contrary to Scripture. Galileo was forced to recant his views. He was more fortunate than his contemporary, the scientist, Bruno, who was burned at the stake for his scientific views. It is now 400 years since the start of the Copernican Revolution. Everyone today (or almost everyone) accepts that the earth rotates along with other planets around the sun. No one would consider this to be a religious matter today and yet at one time it was considered to be a threat to the church. Do you think that planetary rotation is contrary to your religious beliefs?

Over the past 400 years there has been a significant number of observations and interpretations in science that were considered to be inconsistent with Scripture. It became clear that science could not develop as a discipline if it were dominated by Scripture. And so, over the years science has gradually divorced itself from natural philosophy. In his church, the scientist inquires with others concerning the purpose of human existence. In his laboratory, however, he looks for interpretations of the world around him in terms of natural laws.

It will be interesting for you to turn to the books of the history of science and to follow through in detail a scientific concept that shows an interaction with religious authority. It is noteworthy that most of these many scientific ideas that at one time were considered to be heresies are no longer considered to be religious matters and many of these ideas have now become accepted as scientific law.

Aristotle had proposed that greater weights fall the same distance in less time than lesser weights. That is, a 100-pound stone would fall to earth two times as fast as a 50-pound stone. By Galileo's time some scientists knew from observations that this theory was not true, but the writings of Aristotle were held in such high esteem by the church that no one dared question their authority. Galileo recognized and stated that the truth about nature can best be unraveled by disregarding religion. Some of the church leaders at that time pointed out that "the Bible is meant to show people how to go to heaven and not how the heavens go." The same statement was made in 1983 by the present Pope John Paul to a group of scientists.

Aristotle believed that the heavenly bodies revolved in perfect circular paths because "all that is eternal is circular." Over the centuries, however, it became clear to astronomers that planets do not move in absolute circular paths but rather as elipses around the earth. Scientific and religious thought were so intermingled that the idea of circular motion was perpetuated.

The idea was put forth by Newton that the orderly rotation of planets was controlled by gravitational forces. The idea of gravity was attacked as subversive to natural and revealed religion.

Galen (131-201 A.D.) proposed that blood travelled between the left ventricle to the right ventricle of the heart. So great was the prestige of Galen that his books on anatomy were read uncritically for almost 1,500 years. Vesalius (1514-1564) found that the septum of the heart was too thick to allow for blood to ooze between the right and left ventricles. Such was the appeal to authority that Vesalius required courage to make this pronouncement. It would seem to us today that the circulation of blood is strictly a scientific matter.

The use of tissue grafts from one person to another or from a person to himself has been useful in skin burns and in organ transplants. In the sixteenth century, transplantation techniques were described by Gasparo Tagliacozzi, a professor at the University of Bologna. He was bitterly attacked by the clergy at that time for interfering with the handiwork of God. Some religious groups today still hold this view.

A classical argument in biology relates to whether development occurs by means of preformation or by epigenesis. The concept of preformation, proposed in the seventeenth century by Jan Swammerdam, states that development is brought about by the enlargement of a preformed miniature embryo that is found in the germ cells. This concept was extended to assume that these miniature embryos had smaller embryos within them, and in these embryos even smaller ones, etc. This was called "the encasement theory," a concept of boxes within boxes. Biologists of that day thought that they saw a small embryo (homunculus) in the egg or the sperm. Caspar Wolff in the early nineteenth century showed, however, that development occurs by means of epigenesis, the appearance of specialized new structures from more generalized structures.

When we view the concept of preformation today it does not seem to have any bearing on religion. What religious implications do you see? This concept, however, was favored by certain religious groups because it was compatible with the idea of total creation, immortality of the soul, the ultimate fate of man, and to the doctrine of original sin. Preformation fitted in with the doctrine of predestination, in that God had placed a finite number of homunculi within the gametes of Adam or Eve. When the last homunculus had developed, the existence of mankind would end.

When fossils were first discovered they were considered by most people to be accidents of nature, for "to suppose them to be real would require that species could become extinct and that great changes had taken place on the surface of the globe for which there is no evidence in reason, in scripture, in secular history" (Greene, 1959). Most fossils were marine species and an explanation was needed to account for their presence on land far removed from the ocean. This indicated that oceans had once been where land is now located, and that there was continuous change in the earth, in the movements of the seas and the land that could be brought about by the natural causes of wind, water, volcanic fire. Noah's flood was not considered by scientists to be an adequate explanation, because the earth was laid down in layers and different animals were found in different layers. It was recognized also that fossils were different from any known living species. "Why Providence should have destroyed an animal or a species it once thought proper to create is a matter curious inquiry and difficult solution" (Greene, 1959).

The frustration in dealing with the religious constrictions in the development of science was expressed by Robert Hooke, who discovered the cell in 1665 (Greene, 1959).

> Tis a vain thing to make experiments and collect observations, if when we have them, we may not make use of them; if we must not believe our senses, if we may not judge of things by trials and sensible proofs . . . but must remain tied up to the opinions we have received from others, and disbelieve everything tho never so rational, if our received histories doth not confirm them; . . . we should have no more to do but learn what they thought fit to leave us. . . . I hope that sensible evidence and reason may at length prevail against prejudice and at last produce a true philosophy.

It was common in the 1600s to try to explain the world in terms of its purpose. Everything was made by God for a reason. Frequently, this reason related to a service for man. Chickens are here on earth because they give eggs; cows give milk. Mountains exist to bring fresh water to men and animals. John Ray, an eminent biologist, wrote a scientific book as a tribute to the Creator, *The Wisdom of God Manifested in the Works of Creation* (1701). Francesco Redi performed a classic experiment to refute the idea that living things could arise from slime or putrefaction. He showed that houseflies arise only from preexisting houseflies (1688). The language of his scientific paper shows the influence of his religion on his science: "I express my belief that the earth, after having brought forth the first plants and animals at the beginning, by order of the Supreme and Omnipotent Creator, has never since produced any organisms." Today, in scientific publications, there is no reference to the Creator. This is not to deny the existence of a Creator and His work but rather to admit that the concept of the Creator cannot be studied by scientific method. This is a limitation of science, but the impact of religious thinking on science persists into the twentieth century.

Joseph Campbell in his book, *Myths to Live By* (1972), tells about an incident in which he was asked by an Indian, "What are you Western scholars now saying about the dating of the Vedas?" The Vedas are the Scripture for the Hindu. He replied that the Vedas had been written 1500-1000 B.C. and that even in India there was proof of the remains of an earlier civilization than the Vedas. "Yes," said the Indian gentleman, "I know; but as an Orthodox Hindu I cannot believe that there is anything in the universe earlier than the Vedas."

A certain sheik in 1966 asked the King of Saudi Arabia to suppress a heresy that the earth moved around the sun. Those who denied that the earth is fixed and stable were stating falsehoods toward God, the Koran, and the Prophet.

One needs to inquire seriously what impact the restrictions in scientific ideas have made upon the education of citizens and the development of technology. What would be the present state of space exploration if the concepts of planetary rotation and of gravity had not been developed. What would be the present state of medicine if we were not permitted to challenge the ideas that blood moves between the right and left ventricles through the septum. What would be the present state of our health if we denied blood transfusions and smallpox vaccinations to our citizens?

The development of basic scientific concepts eventually leads to a use to man; to enable him to cope with his environment, maintain his health, to maintain a society. When certain religious leaders interfere with the development of basic ideas in science because of apparent contradictions with Scripture, they are not merely playing silly philosophical games, they are interfering with the education of future citizens and scientists; they are destroying the sense of wonder that comes from the discovery of new scientific ideas; they make science appear irrelevant and even obscene. Scripture has not been useful to explain the laws of nature nor to develop scientific concepts.

THE OVERTON DECISION RELATING TO ARKANSAS ACT 590

In 1981, the State Legislature in Arkansas passed into law Act 590 which required that the public schools in the state give balanced treatment to creation science and evolution science. This act was challenged in the federal district court of Eastern Arkansas by the American Civil Liberties Union and 20 other plaintiffs, including clergy. On January 5, 1982 Judge William Ray Overton overturned Arkansas Act 590, stating that "creation science has no scientific merit nor educational value." "Since creation science is not a science, the only real effect of Act 590 is the advancement of religion. Act 590 is an effort to introduce a biblical version of creation into the public schools. This would violate the Constitution's guarantee of separation of church and state."

In the trial, the plaintiffs sought to prove that creation science was a religion, while the defense contended that creation science was a science. Judge Overton contended that "creationism" was an offshoot of fundamentalism. Act 590 defines "creation science" as scientific evidence that indicates: (1) sudden creation of the universe, energy, and life from nothing; (2) the insufficiency of mutation and natural selection to explain evolution; (3) changes only within fixed limits of originally created organisms; (4) separate ancestry for man and apes; (5) explanation of the earth's geology by catastrophism, including the occurrence of a worldwide flood; and (6) a young earth. "Evolution-science" means the scientific evidences for: (1) emergence of life from non-life and order from disorder by natural processes; (2) sufficiency of mutation and natural selection to explain the existence of present-day organisms; (3) common ancestry for man and apes; (4) explanation of the earth's geology by uniformitarianism; and (5) an old earth, billions of years in age.

Creation science is considered to be religion because it is based on the first 11 chapters of the Book of Genesis. The creation week and destruction of the world by flood is found in Genesis.

To teach "creation-science" vs. "evolution-science" is a two-model approach in which one must accept the literal interpretation of Genesis or else believe in the godless system of evolution. To use an "either-or" approach serves no educational or scientific purpose. Although the origins of life involve biology, the scientific community does not consider the origins of life as a part of evolutionary theory (organisms that live today arose from different species of organisms). Contrary to the Act 590, evolution does not presuppose the absence of God.

Science is characterized as: (1) it is guided by natural law; (2) it must be explained by natural law; (3) it is testable; (4) its conclusions are tentative; and (5) it is falsifiable. Creation-science fails to meet these criteria for science. (1) There is no scientific definition of "kinds." It is not known how many kinds were created. (2) Separate ancestry for man and apes refers to no scientific fact or theory. (3) The existence of a worldwide flood or of geological formation by catastrophe is not based on scientific evidence but rather on the Noachian Flood. (4) A world of "relatively recent inception" has no scientific meaning. It is based on an estimate of 6,000-10,000 years by use of genologies of the Old Testament.

Articles espousing creation-science theory have not been published in recognized scientific journals. The creationist methods do not involve the analysis of data. They take the literal wording of Genesis and try to find scientific support for it. The Creation Research Society requires that members subscribe to the belief that Genesis is historically and scientifically true. One cannot have a science if one starts with a conclusion and refuses to change it regardless of evidence.

It is appropriate and desirable for everyone who has a knowledge of evolution to criticize the evidence for this concept. This is the way that science grows. It is necessary to recognize, however, that there are two separate activities. One is scientific; the other is educational. Creationists need to interact more with evolutionists at the level of scientific dialogue. They need to present scientific evidence for creationism and not merely evidence against evolution. They are badly in need of a substantial increase in trained and active research scientists who hold the creationist point of view. They need to publish evidence in scientific journals. They need to try to falsify the theory of creationism by means of observations and experiments. There appears to be a lack of self-critical analysis in their work. Professional ministers with little training or experience in mathematics, chemistry, biology, and geology are inadequate to discuss scientific experiments. Emotion is no substitute for evidence. The only way that creationists can change the educational system in a democratic society is by showing that they are correct in their science. Creationists can no more prevent research into evolution and the teaching of evolution in the science classes of *public* institutions than they can prevent the teaching of planetary rotation. For creationism to be taught as a science requires that it be explainable by natural law.

Levels of Rejection and Acceptance of the Concept of Evolution

People accept or reject the concept of evolution for different reasons.

1. Evolution is unacceptable because it is an atheistic and immoral idea.
2. Evolution is unacceptable because it is contrary to the biblical stories of Creation and the Flood.
3. Evolution is unacceptable, but the "old earth concept" is acceptable. A "day" does not have to mean 24 hours. When we speak of "a day of judgment," we do not necessarily mean 24 hours. A "day" could be millions of years.
4. Evolution is acceptable within limits. In the beginning God created "kinds." These kinds are not related by common ancestry. The variation of kinds living today from a single originally created kind has occurred. If you want to call this evolution, it is all right.
5. Once organisms were created, evolution could have occurred. Organisms, however, could not have arisen without the intervention of God. Life could not have arisen by means of the synthesis of large molecules. Complex and functional proteins could not have been formed by chance.
6. The critical question is not organismic evolution but rather the origin of life. The major question is not evolution but rather when did God intervene. Did God make Adam or Atom?
7. Evolution is acceptable, because the sequence of events listed by scientists is like the sequence of events proposed in the Bible.
8. Evolution is acceptable, but the process of natural selection is not acceptable. Natural selection occurs by chance and therefore lacks purpose. God did not create and plan a world based on chance.
9. Evolution is acceptable provided that it is a teleological process proceeding to a goal. The goal of evolution is the creation of man.
10. Evolution is acceptable because of the scientific evidence that has been accumulated for 125 years. The views of religion are irrelevant in scientific matters. Evolution is not a religious matter.

Most Christian sects are opposed to the teaching of Creationism alongside of evolution in public school science classes. They are fearful that many people will be alienated from religion and that the Bible will be analyzed as a scientific text. An interesting and informative paper, "Creationism and Evolution, The Real Issues," was published in 1981 by N. Patrick Murray, an Episcopal priest, and Neil D. Buffaloe, a biologist. These men are citizens of Arkansas. They point out that the issue is not the validity of the scientific concepts. Further, the issue is not faith in God. Many theists accept the concept of evolution. God created the universe and how he did it is not a challenge to their theistic belief. The central issue is biblical interpretation. "Scientific creationism" rests on biblical literalism. Biblical writings are divinely inspired and are true in all statements whether they relate to morals, history, or science. To scientists, the concept of evolution is religiously neutral. It describes a process. It is not a threat to theism.

A position paper of the United Church Board for Homeland Ministries was published in 1983. This paper "Creationism, The Church, and The Public School" stresses the importance of the public school to both the church and society in preparing children to live in the culture of the twenty-first century world. The public school is the cornerstone of our democratic society and must not be answerable to religious groups. This paper opposes Arkansas Act 590. It affirms that belief in God does not depend on scientific discovery; that evolution is the best present-day scientific explanation of life; that secular education is based on the efforts of the academic community; that religious commitment is the responsibility of the home and church, not of the public school; that scientific textbooks should not exclude the concept of evolution; that belief in a Creator God and evolutionary theory are compatible.

In the *New York Times* (11/14/83) there is a news statement by Philip M. Boffey, entitled *"Once-Hostile Vatican Forges Close Links with Scientists."* The article reports on a meeting of scholars, of varying religious beliefs, that was convened by the Vatican. These scholars concluded that the concept of human evolution is supported by masses of evidence from paleontology and molecular biology beyond serious dispute. This theory was opposed by some Catholic theologians in the last century.

The journal *Science* (1982, Vol. 215, 142-146) has an excerpt from the creationism trial in Little Rock, Arkansas, *"Where is the Science in Creation Science."* This article, by Roger Lewin, is a good one to read and discuss. It is taken from some of the verbatim testimony from the trial and inquiries into the competency of the creation scientists from their testimony.

An excerpt of Judge Overton's decision has appeared in *Science* (1982, Vol. 215, 381-384) by Roger Lewin.

The entire trial has been published in *Science* (215: 934-943) in 1982.

The Galileo Controversy

Authors frequently have a way of saying things that get to the core of the argument. This is shown by Eric Bently in his play, *The Recantation of Galileo Galilei*. You will enjoy this work. I think that the Galileo controversy is central to the evolution controversy and to all science-religion controversies. Bently makes the following points. There is the matter of defending the faith. What is important is personal salvation and the worth of mankind. The stars are so far away that we might just as well ignore them. It saves us the bother of disturbing our intellect and enables us to turn to the more important matter of our souls. We must not ask the question whether the Christian story can be fitted into the facts. Suppose that there are an infinite number of worlds. Did God send out an infinite number of sons or the same son an infinite number of times? Many discoveries are too much for people, and such terrible things are going to be discovered.

It is startling and educational to read Galileo's Recantation:

> I, Galileo Galilei, do now, with a sincere heart and unfeigned faith, abjure, detest, and curse the heretical belief that the earth moves around the sun, and take my oath that never again will I speak, write, or otherwise assert anything that might lend plausibility to this belief. Should anyone suspected of subscribing to this heretical belief be known to me, I shall denounce him to this holy office in Rome or to the inquisitor in whatsoever place I shall be, so help me God and this holy book.

It is now 350 years since Galileo. Despite this and many less famous incidents, the science-Scripture battle continues. I am sure that the Catholic church does not think that the sun moves around the earth. This incident is still a matter of embarrassment to the church.

On July 1, 1968 it was announced that the Roman Catholic church might revise its censure of Galileo. It was stated that such a revision "could heal one of the deepest wounds between science and religion" (*The New York Times,* July 2, 1968).

On May 9, 1983 Pope John Paul II said that the Vatican was still examining the "heresies" of Galileo. This experience with the Galileo affair led the church to a more proper attitude and a more accurate grasp of its own authority. "It is only through humble and assiduous study that she learns to dissociate the essentials of faith from the scientific systems of a given age" (*The New York Times,* May 10, 1983).

Do you think that evolutionists should recant?

It may be instructive to look broadly at the relation of science and religion through the eyes of scholars who are involved in the study and analysis of the Bible and in the origin and history of religion. Toy (1924) in *Introduction to the History of Religions* makes the following comparison. Both science and religion inquire into the nature of the universe. Both scientific and religious concepts change and are refined through human experience. Science and religion deal with different aspects of human inquiry, and therefore there should be no permanent hostility between them. Religion seeks the ultimate moral ground and the Prime Mover of the universe. Science looks for natural causes to explain the universe.

Science and religion clash when a concept considered to be a religious belief is challenged by a scientific discovery. The religious establishment initially protests and considers the new scientific concept to be heretical. Later, however, it accepts a well-attested scientific conclusion and brings it into harmony with its fundamental beliefs. The religion continues to survive and to flourish. It is not destroyed by the scientific concept. There are notable examples of this acceptance of previously "heretical" scientific views: the theory of planetary rotation, the theory of gravity, the magnitude of the stellar universe, the great age of the universe, the theory of evolution, the theory of the conservation of energy.

Scientific investigations thus have delimited religion. Although religious explanation is influential and important, it cannot include the criticism of scientific theories in its domain. This divorce from scientific activity must ultimately be favorable to the development of religion. It certainly saves it from embarrassment and from credibility loss.

Objectivity in Teaching

Mrs. Jones applied for a position to teach science to public school students. She appeared before the school board for an interview. She was asked, "Mrs. Jones, do you believe that the sun goes around the earth or that the earth goes around the sun?" Mrs. Jones really needed the job. She said, "I can teach it either way."

CHAPTER 6

THE GENE CONCEPT

THE GENE CONCEPT is treated here as a scientific idea that has moved from speculation to principle. The increase in confidence in this concept has involved 100 years of work involving thousands of scientists and hundreds of thousands of scientific papers. It has involved different kinds of organisms (not just garden peas). It involves a great amount of self-critical analysis by trained and devoted scientists. A person who teaches or studies science must study the historical development of some scientific concept in order to understand how the scientific community operates and how theories are accepted or rejected.

* * * *

The science of heredity is based upon the concept that there are units of inheritance (genes). The idea of hereditary particles that maintained their properties and could not be diluted out is comparable to the idea of the atom in chemistry as a basic unit of matter. How do we know that there are genes? How does the community of biologists make up its mind that genes exist? Why should we believe them?

Before the nineteenth century, it was recognized that offspring bore resemblances to parents. By means of breeding towards the desired type, new varieties of domestic animals had been developed. These breeds of useful animals were developed by people who did not understand the nature of hereditary material. They thought there was a substance that circulated in the blood and somehow was passed on to the next generation. This substance could be increased in amount or could be diluted out. This would result in gradual changes in appearance from one generation to the other.

The experiments and interpretation leading to the concept of the gene were done by Gregor Mendel, a priest in a monastery in Moravia (now Czechoslovakia). Mendel did his work on the garden pea *(Pisum sativum)*. He selected seven pairs of contrasting traits for study (round vs. wrinkled seeds, yellow vs. green seeds, long stem vs. short stem, white flowers vs. colored flowers, etc.). He made crosses between each pair of true-breeding plants and obtained comparable results for each.

One set of crosses by Mendel involved plants with round seeds and plants with wrinkled seeds as the contrasting traits.

P_1 round x wrinkled
F_1 round (interbred or self-fertilized)
F_2 round wrinkled
5,474 : 1,850
2.96 : 1
3 : 1

Mendel crossed plants with the round trait to those with the wrinkled trait in the P_1 (first parental generation. From this cross he obtained plants in the F_1 (first filial generation) that produced only round seeds. He allowed these F_1 plants to interbreed or to self-fertilize, since each pea plant produced both eggs and pollen grains. From such a cross there was produced in the F_2 (second parental generation) both round and wrinkled seeded progeny. From counts of the number of each type, 5,474 round, to 1,850 wrinkled, Mendel inferred the 3:1 round to wrinkled ratio from these data. This 3:1 ratio could explain all of the seven crosses involving different pairs of alternate traits.

It is clear from these data that Mendel was not dealing with a "blending type" of inheritence. The traits appearing were clearly round or wrinkled and not intermediate in appearance. Although the wrinkled trait had been lost in the F_1, it reappeared, and from a round parent, in the F_2. Mendel explained his data by assuming that there were factors (genes) for the round (W) and wrinkled (w) traits.

P_1	WW		x	ww
gametes	W			w
F_1			Ww	
gametes	eggs			pollen
	W w			W w
F_2	WW	Ww	Ww	ww

Mendel assumed from his data that factors (genes) were passed on from one generation to the next by means of eggs and pollen. These factors occurred in pairs in each plant (either WW, Ww, or ww). In eggs and pollen, only one factor (W or w) appears. Thus, in the cross above, the round (WW) plant produced only gametes with the W gene, and the wrinkled (ww) plant produced only gametes with the w gene. These gametes (egg and pollen) came together in fertilization to produce the Ww fertilized egg which developed into the plant with round seeds. The F_1 plant (Ww) would now produce two types of eggs and pollen in equal numbers, those bearing the W factor and others bearing the w factor. These could combine in fertilization to yield round (WW, Ww) and wrinkled (ww) progeny.

The explanation of the 3:1 ratio is important, since it points to the element of chance that is involved. Notice that the 1WW:2Ww:1ww is also a ratio predictable by chance. Take two pennies, toss them simultaneously into the air and recognize that they appear in three groups (head-heads, heads-tails, tails-tails). Predict the ratio that you would expect to get. Now try the experiment. Toss these two pennies 100 times and record your data. What ratio do you get?

As a further test of the segregation of Mendel's factors, predict the results that you would expect from a cross involving an F_1 plant with a wrinkled plant, and of a wrinkled plant with a wrinkled plant.

Gregor Mendel is recognized as the discoverer of the units of heredity (now called genes). Did Mendel actually see genes? No! What Mendel saw was the segregation of traits (round, wrinkled). He saw that the wrinkled trait appearing in progeny from Ww (round) parents was not diluted out. From the segregation of traits, Mendel inferred that there was a segregation of genes that controlled these traits. What Mendel did was to suggest an hypothesis to explain the segregation of traits. His hypothesis is a scientific hypothesis, in that it is testable (falsifiable) and one can make predictions from experiments. The "gene theory" of Gregor Mendel was a speculation. Do genes really exist as chemical substances, or are they fantasies in the minds of biologists? Mendel's proof of the existence of genes was based on circumstantial evidence. At one time it was questioned whether atoms actually existed or whether the atom was only a useful concept. If the scientist relied only on direct evidence, we would know very little about the universe.

How were Mendel's experiments received when published in 1866? No one recognized their significance, nor that this was one of the great concepts in biology. Mendel was ahead of his time. He never knew that he had made an important contribution to science. To appreciate Mendel it was first necessary to discover chromosomes and to recognize their significance. In 1866, the role of chromosomes in mitosis, meiosis, and fertilization was not known. It was during the period of 1870-1900 that chromosomes were discovered and their role in reproduction was studied. It was only then that it could be hypothesized that chromosomes carry hereditary material. In 1900, three biologists each independently and each unaware of Mendel's work repeated Mendel's experiments on other organisms and obtained Mendelian results. It is poignant that if Mendel had never existed, we would still have had the concept of the gene. Darwin was lucky. He came at the right time. Even if Darwin had never existed, we would have nevertheless the concept of evolution as a part of biology because other investigators independently had made this discovery! In the universe there are laws of nature waiting to be discovered, if not by one scientist, then by another. In 1900, Mendelian inheritance emerged as an attractive hypothesis that needed to be tested and challenged.

Many biologists did not accept the interpretation put forth by Mendel or thought that his experiments dealt with some marginally important aspect of biology. This is the way it is with new scientific ideas. New concepts do not stand out as truths to everyone. It takes a certain incubation time and a time for evaluation and experimentation. Biologists repeated Mendel's experiments on other organisms. Mendelian data and interpretation were found to hold and to have predictive value for a wide range of plants and animals and microorganisms. Mendel's laws gradually became accepted as basic to the transmission of hereditary traits. They became a principle upon which one could build. By 1920, the existence of genes was universally accepted. No one however had ever seen a gene nor knew of what chemical substance (or substances) it was made of.

Mendel's principles have been important in the field of medical genetics. Analysis of pedigrees has enabled genetic counselors to predict the chances that certain types of progeny could arise. Mendel's paper dealing with garden peas is probably the most important paper in human genetics. Mendel's research was basic research, in that it took some relatively insignificant traits (round and wrinkled seeds) and developed a biological principle from them. This principle is applicable to the practical research involved in understanding human disease.

The Localization of Genes on Chromosomes

Late in the nineteenth century, it was suggested that hereditary material must be localized in chromosomes. Indeed, when gene segregation was rediscovered in 1900, chromosomal segregation had already been established. It was known from studies of fertilization in invertebrates that the egg and sperm fused to form a zygote, which in turn developed into a new adult. The egg was large and the sperm relatively small. The nucleus of each gamete, however, was the same size. Since each parent contributed equally to the offspring, it was suggested that male and female nuclei contained equivalent hereditary material. In the process of fertilization there is a fusion of the male and female nuclei to form a single nucleus. Studies of cell division and gametogenesis revealed the allocation of chromosomes into daughter cells and an understanding of the alternation between haploid and diploid states in the life cycle. In grasshoppers, a difference in the sex phenotype could be correlated with the fact that females contained one more chromosome than males, and it was speculated that this extra chromosome was a sex determinant. After 1900, it was established from the study of many kinds of organisms that the number and array of chromosomes is characteristic for each species.

After 1900, the study of chromosomes by direct examination under the microscope (cytology) and the study of inheritance by means of breeding techniques (genetics) fused to form the field of cytogenetics. Cytogenetics was concerned with elucidating the relation of genes to chromosomes.

Biologists recognized that there was a parallelism in segregation behavior between genes and chromosomes that could only be explained by the hypothesis that genes are located on chromosomes (the chromosomal theory of heredity).

Chromosomes occur in pairs. The chromosome number is diploid for somatic (body) cells but is haploid for mature eggs and sperm (germ cells). Fertilization results in a doubling of the chromosome number (1N → 2N) and gametogenesis in a halving of the chromosome number (2N → 1N). These observations are consistent with the concepts regarding the segregation of genes.

Unusual instances of gene segregation could be explained by unusual chromosomal segregation. This was shown first in the case of sex linkage for white eye color in *Drosophila.* When red-eyed females were crossed to white-eyed males, their progeny were all red-eyed (males and females). When, however, white-eyed females were crossed with red-eyed males, their progeny were red-eyed females and white-eyed males. There was a phenotypic difference related to sex type. The explanation of this segregation was related to the presence of this gene on the X-chromosome in a scheme where females are XX and males are XY. Many instances of this kind of inheritance have been shown for man. Indeed, whenever we find this unusual kind of inheritance for a newly discovered gene, we know that this gene occupies a position on the X-chromosome.

It frequently happens in science that an exception arises that proves the rule. Calvin Bridges' work on the inheritance of white eye color in *Drosophila* was published in 1916 and was important in establishing that genes are located on chromosomes. Bridges crossed white-eyed females to red-eyed males and obtained not only the expected red-eyed female and white-eyed male progeny but also a few exceptional types of white-eyed females and red-eyed males. Bridges explained the occurrence of these exceptional types of proposing the principle of chromosomal non-disjunction. Thus, although ww females normally produce X (w) eggs because the 2 X-chromosomes disjoin (separate), occasionally XX (ww) eggs and eggs without an X (or w) are formed due to the X chromosomal non-disjunction.

Under this particular scheme, when one examines the cells of the white-eyed females one finds three sex chromosomes (XXY), and of the red-eyed males, one sex chromosome (X). This experiment by Bridges has been diagrammed many times. Look it up in almost every genetics text.

We have here again an experiment in which the unusual transmission of genes can be explained by the unusual transmission of chromosomes. There is a correlation between gene inheritance and chromosomal inheritance. This is circumstantial evidence. We still do not see the genes. On the other hand, there appears to be no other way to explain these observations.

In science there is often a practical application to basic research, but there is usually a lag in time. Few people care whether *Drosophila* have white eyes or red eyes, but we do care whether people have Down syndrome. The principle of non-disjunction discovered in 1916 by Calvin Bridges working on *Drosophila* explained, in 1959, the cause of Down syndrome.

Some gene combinations assort at random to yield gametes in equal numbers, while other gene combinations tend to remain together. Assume that the gene combination AaBb yield four kinds of gametes (AB, Ab, aB, ab) in equal numbers, and that the gene combination AaCc yields four kinds of gametes where two (AC, ac) are higher in frequency and the other two (Ac, aC) are lower in frequency. In those instances where random assortment holds, the genes (in this case A,a vs. B,b) are located on non-homologous chromosomes, while A,a vs. C,c are linked, that is, located on the same kind of chromosomes. Can you test this hypothesis by determining the gametic frequencies for BbCc? What data would you expect under the hypothesis of linkage? The idea of linkage led to the placing of genes at specific loci on specific chromosomes. The more frequently that gene combinations remained together, the closer they were located on the same chromosome. We know the positions of genes on chromosomes even though we had never seen genes or knew what they were. Look up the linkage maps for the genes of *Drosophila* and of man. Notice that the genes occupy definite positions relative to one another.

The position of genes on chromosomes could be determined also by deficiencies. The salivary gland chromosomes found in *Drosophila* enable one to show an elegant correlation between genes and chromosomes. These chromosomes are large and show a transverse banding pattern. There are about 5,000 such bands on the *Drosophila* chromosomes. These bands are characteristic in their form for each chromosome region. It has been demonstrated that when certain bands are absent from the chromosomes, certain traits do not appear. A chromosomal deficiency results in the absence of the gene. This is shown, for example, for the X-linked gene the white locus in *Drosophila.* Similar experiments involving deficiency in bands and other parts of the X-chromosomes also showed that other genes did not express themselves. Thus, one can demonstrate that genes are located on chromosomes and that each gene has its characteristic locus.

By 1940, the concept that genes existed as real entities and that they occupied specific loci on specific chromosomes was accepted by most biologists. The chromosomal theory of heredity had been tested on many different species and found to hold. For my own part, I was able to demonstrate the validity of this theory for my own experimental organism, a parasitic wasp, *Habrobracon juglandis.* I was able to build upon this theory and make predictions from it. At what point should the individual biologist believe that there are genes and that they are located on chromosomes? This is an individual decision based on examination of the evidence and a faith in the integrity of one's colleagues. There is no council of ordained biologists to say "you must accept the chromosomal theory of heredity." No member of Genetics Society of America is required to sign an oath stating that he believes in the chromosomal theory of heredity. The chromosomal theory of heredity has become a principle. But it was not until after 1950 that it was shown that genes are made up of a chemical substance, deoxyribose nucleic acid, DNA. And yet, it was during this period (1930-1960) that the genetics controversy took place in the Soviet Union and the existence of genes was ruled out by politics. Unlike the Soviet Union, in most countries, the concept of the gene was not a political matter.

Although the concept that genes occupy definite positions on chromosomes was accepted, there were other observations that seem contrary to it. It has been recognized that genes occasionally occupy different positions on chromosomes and may come to lie adjacent to a different group of genes. When this occurs there is often a change in the expression of a trait. It has been established that in the development of a corn plant from seed that certain genes regularly move about from one part of a chromosome to another part or to a different chromosome. This phenomenon of "jumping genes" has been found also in microorganisms and is also important in the action of genes that cause cancer. In December 1983, Barbara McClintock was awarded the Nobel Prize in Physiology and Medicine for discovering "jumping genes." It remains to be found under what conditions the genes can move about and insert themselves into new positions and under what conditions these genes remain in fixed positions on the chromosome. In science one discovery is often dependent upon another. The discovery of "jumping genes" could not have taken place before the discovery that genes occupy specific loci.

The Gene as a Chemical Substance

Genes are made up of a chemical called deoxyribose nucleic acid (DNA). How do we know this? It was known from the study of genetics that genes are located in chromosomes, that chromosomes are found in the nuclei of cells, and that the chromosome number for somatic cells is constant (2N) and double the number for mature gametes (1N). It was reasoned from this that the chemical that was genic must also be located in chromosomes, be constant in amount from somatic cell to somatic cell, and be half in amount in mature gametes. DNA, but no other chemical, satisfied these conditions. In 1945 it was demonstrated by the biologist Avery and colleagues that DNA, extracted from a virulent strain of pneumococcus (a bacterium that causes pneumonia) and placed into a test tube with a living, non-virulent strain, could transform some of the non-virulent pneumococci into virulent types and that these newly transformed virulent cells continued to produce virulent progeny. This experiment should have convinced the biologists that DNA was the genetic material. At the time, however, biologists had a mistaken view of the chemical nature of DNA and considered it to be too simple a substance to be hereditary material. Indeed, it did not occur to them to think that DNA could be the hereditary material. They thought that only proteins were sufficiently complicated to account for the considerable amount of gene variation. But biochemical work in the late 1940s showed that DNA was also complex, and in 1952 it was shown for the inheritance of viruses that viral DNA was passed on from one generation to the next, but viral protein was discarded. Because of these experiments, biologists by 1952 were convinced that DNA is the genetic material. This resulted in a strong stimulus to discover the chemical nature of DNA.

James Watson, an American virologist and Francis Crick, a British chemist, determined the structure of DNA in 1953. Their work ranks as one of the great discoveries in science. The elucidation of DNA structure stimulated many researches by others and led to the development of the field of molecular biology. It led to a precise definition of the chemical nature of the gene, to an explanation of how proteins are made, to an understanding of gene replication and gene mutation. It has led to a DNA technology, in that genes could be made in test tubes. A large biochemical industry is developing whereby useful proteins such as insulin and growth hormones can now be made from genes rather than being collected from the organs of slaughtered animals.

In this example of scientific discovery we have progressed from the speculation that there exist certain units of heredity that we name "genes" to the ability to make genes in the test tube. This scientific accomplishment has required the work of a community of scholars working together for a period of one hundred years.

* * * *

Picture Frames

How does the biologist know that there is a process of cell division that he calls mitosis? At one time (about 1830) it was thought that the cell nucleus was a small embryonic cell developing within a larger cell.

Starting in 1870, certain techniques were developed that enabled one to study cell division under the microscope, and thus the field of cytology (study of cells) developed. It was discovered that certain dyes would stain the cell; some dyes stained the nucleus and others the cytoplasm. Chromosomes (colored bodies) were discovered and so named because they had an affinity for certain dyes. Techniques were developed for the preserving of tissue by certain chemicals, for dehydration, for embedding tissue in paraffin, for the slicing of tissue into ribbons of 10-micron thickness, and for placing these slices between bits of glass for examination under the microscope. During this period the quality of the microscope was improved. Techniques were also developed in which individual cells were dispersed by a "squash technique" and chromosomes could be studied after the application of nuclear stains.

If one places the bottom of an onion bulb into water, within a few days roots will grow. One can cut off the tips of these roots, place them on a glass slide with preservative-dye (such as aceto-carmine) and gently break the tissue apart, thereby separating the cells from one another. Or, one can slice the root into thin ribbons and thereby examine the cells in their original spatial relations. Some of these cells show a nucleus and others show chromosomes. Chromosomes in different cells show differences in length and position. The biologist in examining these dead cells under the microscope does not see the movement of chromosomes. He does not see a metaphase cell change into an anaphase cell. But it is from the study of such dead cells that the dynamic process of mitosis was elucidated. It is much like the putting together of a motion picture from thousands of still pictures. It was not until the development of the phase contrast microscope that the details of mitosis were actually observed in living cells. The processes of spermatogenesis, oogenesis, and fertilization were explained from the study of preserved material. It frequently occurs that a given process cannot be directly observed, and special techniques need to be developed for its study. New techniques give us new ways of "seeing."

What Does it Mean to See?

Seeing involves not only the eyes but also the other senses and the brain. It also requires a mind that is able to plan experiments that enhance our ability to observe. It calls for a mind that is able to recognize order in the universe and to build hypotheses from this order. It requires a society that permits people to challenge and to innovate.

CHAPTER 7

THE GENETICS CONTROVERSY IN THE SOVIET UNION

A RATHER curious incident relating to the mechanisms of inheritance occurred in the Soviet Union between 1925 and 1965. The Russians threw out Mendelian genetics from their research laboratories, their medical institutes, their agricultural institutes, and their schools. At a scientific level there was a controversy whether or not genes and chromosomes exist as hereditary units. At a practical level there was a question concerning what is the best way to improve domestic plants and animals, by gene selection for desirable traits or by the selection of traits that have been acquired by environmental experiences. At a philosophical level what was at stake was the compatibility of a particular scientific view with the Soviet concept of dialectical materialism. The anti-Mendelian biologists were led by an agronomist, Trofim Lysenko. In the 1940s the influence of the Mendelian geneticists became less and finally nil. The Lysenkoists had the upper hand during the premierships of Stalin and Krushchev and held it until Brezhnev came into power in 1965.

The Soviet objection to genetics was influenced by the tenets of dialectical materialism: there is a materialistic explanation of the universe and the origin of man; the present state of man and matter can be explained by natural causes; it is not necessary to invoke a spirit or deity; matter is continuously changing; man and society are continuously changing; these changes are brought about by an interaction with the environment and result in improvement (progress); when change ceases, life ceases.

The Soviet objection to Mendelian genetics is based upon the apparent conflict with dialectical materialism. Soviets claimed that Western genetics was undialectical for the following reasons:

1. Genes are controlling units that are isolated from the rest of the cell and do not interact with the cell. There is no cytoplasmic influence on genes.
2. Genes are stable units. They do not change or changes are so infrequent as to be inconsequential.
3. Genes are isolated from the external environment and thus do not interact with the environment.
4. Even if genes did mutate infrequently, they do so without environmental influence and in most cases the new mutants are non-adaptive (most are harmful and even lethal). How then can one get improvement by mutation?
5. The entire organism operates and interacts as a unit. It cannot be divided into two parts, a hereditary and a non-hereditary part. The organism cannot be dualistic, with one part isolated from the other.
6. Most mutations occur spontaneously; occur by chance. To believe that an event occurred in this way is to deny that there are natural causes for the events. Chance is metaphysical; it offers no explanation. The approach to the improvement of plants and animals by selection of mutations caused by chance is scientifically sterile.

It is clear that because of the requirement for conformity to dialectical materialism, science in the Soviet Union is a political matter. This interaction between science and politics is shown by Zhores Medvedev in *The Rise and Fall of T.D. Lysenko* (1971).

Medvedev points out that in 1937 the debate involving genetics was removed from science and into politics. Geneticists were considered to be "enemies of the people." The scientists who believed that genes exist were considered to be capitalistic and undialectical. It was pointed out that Mendel was a priest and therefore was insincere and badly motivated. The Soviet philosopher Prezent coupled Mendelism with racism and pseudoscience. "The struggle against the remnants of bourgeois opinions in science, the implacable struggle against pseudoscience and the idealistic and metaphysical distortions is the business of every scientist and every scientific institution in our land." Lysenko stated that "Marxism is the only science. Darwinism is only a part; the real theory of knowledge of the world was given by Marx, Engels, and Lenin. . . ."

The political aspects of the controversy were enhanced in reaction to the Nazi view of a master race based on genetic superiority.

Mendelism

Lysenko assumed that there was no special hereditary material localized in the chromosomes. He rejected the theory of the gene on philosophical considerations and on his interpretation of the data collected by geneticists on a wide range of experimental organisms. He stated that Mendel's Pea Laws do not hold; that the 3:1 ratios, the 9:3:3:1 ratios are not true even for Mendel. The experiments by Mendel were repeated by certain Russian biologists who reported that the data indeed do not corroborate Mendel. Analysis of these data by others, however, showed that the data did indeed corroborate Mendel. The Russians, because of their disregard for the laws of chance, had rejected the tool of statistical analysis. Lysenkoists pointed to experiments in which 3:1 ratios were not obtained, involving polygenic inheritance and linkage, concepts that have very nicely been incorporated into gene theory. They tended to quote the older 1910-1920 experiments and not the modern ones. Had they studied the literature dealing with the bread mold Neurospora, they would have found very precise segregation ratios. They also pointed to experiments dealing with the inheritance of cytoplasmic particles (mitochondria, chloroplasts) and thus to show that the cell nucleus is not unique in containing hereditary material.

It is curious that the argument dealt not only with interpretation but also with the validity of the data. Geneticists throughout the world have not only corroborated Mendelism and demonstrated the existence of gene mutations but have built upon these concepts to improve domestic plants and animals and to improve the health of human populations.

Phasic Development

Lysenko returned to the Lamarckian scheme of the inheritance of acquired characters: When organisms are exposed to changed environmental conditions during a "sensitive stage," not only will these exposed organisms be changed, but their progeny will show those same changes even without further environmental stimulus. Lysenko reported that he changed and shortened the growing season for wheat by means of soaking the wheat in cold water (vernalization). By shaking up the "heredity" of the wheat seeds by means of a temperature shock during a specific sensitive phase of development, he reported a permanent change from winter wheat into spring wheat. Experiments have been done by Western geneticists on *Drosophila* and other organisms where temperature shocks during critical periods of development changed certain traits in adults, but these traits are not perpetuated through gametes. In his experiments, Lysenko did not rule out the possibility of contamination of the soil by other strains or genetic variability in his strains. Indeed, if one starts with the view that genes do not exist, then there is no need to consider genetic variability.

Lysenko's followers reported on the transformation of one species into another, wheat into rye, pines into firs. These experiments were fraudulent and support for these mythical transformations was based on the authority of Stalin (Medvedev, 1971).

Vegetative Hybridization

Lysenko reported that when grafts between fruit trees were made, seeds were obtained that gave rise to trees that showed hybrid traits. If branches from an apple tree that bore red fruit were grafted onto an apple tree that bore yellow fruit, *seeds* from the graft would produce trees with mixed fruit of yellow or red, or striped red and yellow apples. The same data were reported for other kinds of fruit trees. Such observations had never been made by horticulturists who had made grafts of thousands of trees.

In all, the experiments of Lysenko have not been corroborated by biologists outside of the Soviet Union. In the case of Mendelism the data collected throughout the world on many different species support the concept of gene segregation and random assortment. The phasic development experiments appear, according to Medvedev, to be fraudulent; the grafting experiments cannot be repeated in other laboratories.

The Genetics Congress

In 1948, at the Soviet Congress on Genetics a debate was held between those geneticists who defended classical gene theory and those who denied the existence of genes as a special kind of hereditary material. There were clear political overtones in the debate, because each side seemed to be interested not only in the criticism of data and experimental methods but in proving that their views of hereditary fitted in with the prevailing Soviet philosophy. Indeed, it was not clear to all Marxists that Mendelism is undialectical. In 1943, a French biologist, Marcel Prenant, wrote a book, entitled *Biology and Marxism,* in which he tried to show that Mendelian genetics is consistent with dialectical materialism. Prenant's loyalty is shown by the following quote from his book. "If this is a good book it is because it is a Marxist book. If it is a bad book it is because it is not sufficient Marxist." When Lysenko's experiments were challenged he appealed to the authority of Marx and Engels and to a plant breeder, Michurin, who worked at the turn of the century. He accused his opponents of being insincere and of heresy to the communistic cause. He accused the gene theory and the chromosomal theory of heredity of being promulgated by religion (Gregor Mendel was a priest) and by capitalism.

The debate ended abruptly when Lysenko read a letter from the Central Committee of the Communist Party.

> Lysenko—Comrades! Before proceeding to the concluding remarks I consider it my duty to declare the following. I have been asked in a memorandum as to the attitude of the Central Committee concerning my paper. I answer—the Central Committee has examined my report and has approved it.

This was followed by a stormy applause and ovation. It meant that the Lysenko view of heredity had become the party line that all biologists must now accept. One can imagine what would happen if the Congress of the United States were to pass a resolution that all biologists must accept the chromosomal theory of heredity.

The Soviet scientists got the message, loud and clear. There followed pledges of loyalty and support of the Lysenko views and recantations of Mendelian views. Here are a few recantations taken from the Proceedings of the Lenin Academy (1948).

> From tomorrow on, I shall begin to rid not only my own scientific activity of old reactionary views, but will also begin to remake all of my students and comrades.
>
> My speech of two days ago was unworthy of a Soviet scientist and a member of the Communist Party. . . . I declare here that I will wage this war and give it political significance. Only the alien scientist who will understand that the bridge must be crossed towards us and not towards him may depend on our attention. Believe me, that today, I make a Party step and appear as a true member of the Party.
>
> I consider it to be my moral duty to be a sincere Michurinist, a sincere Soviet biologist.

These recantations after the decree by the Central Committee show that this is not a scientific but rather a political matter. Julian Huxley (1949) summed up the Lysenko controversy:

> The issue could not be stated more clearly. Do we want science to continue as a free pursuit of knowledge and of control of nature or do we want it to become subordinate to political theory and the slave of national governments? It is a crucial question, on which the general public as well as the professional scientist must make up its mind.

Impact on Soviet Science and Education

What was the impact of this political decision on Soviet science and education? The best scientific minds were prevented from carrying on research. Many biologists were removed from positions of influence. Some were placed into prison. There was the loss of a generation of scientists. In medicine, the institutes of medical genetics were closed. Since there are no genes, there could be no genetic disease. The genetic basis for blood groups were denied. In agriculture, the institutes of animal science that dealt with lethal mutations in domestic animals were closed. Since there were no genes, there could be no deaths due to lethal mutations. In the plant sciences the collections of seeds from different parts of the world were discarded. Methods of plant improvement based on breeding and selection were replaced by trying to improve plant lines by environmental stimuli. The Soviets injured their agriculture. In education, books on Mendelian genetics were purged from the curriculum. Books and scientific journals from other countries were carefully censored and certain papers were deleted. Soviet scientists did not attend international scientific meetings. In short, communication with the rest of the world in scientific matters was lost.

The purge by Lysenkoists not only discarded Mendelian genetics but also other landmarks in science. And they began by declaring reactionary and idealistic Einstein's relativity theory ("reactionary Einsteinism"), Bohr's complementary principle, Pauling's theory of resonance. The repressiveness extended to music, art, and literature.

It is very difficult to free one's scientific convictions from one's political convictions. An entire generation of Soviet scientists carried on research along Lamarckian lines. Even after the discovery, in 1952, that DNA was the hereditary material, Soviets continued to deny the existence of genes! In the United States many communistic sympathizers sided with Lysenko. From many such personal arguments I have found that one cannot by scientific reason alone change the minds of people with strong political, philosophical, religious conviction. This is particularly true if they do not do the hard work of learning the science.

Meeting of the Genetics Society of America (1949)

The 1948 Soviet Congress on Genetics had considerable impact on citizens in the West. How could a country that had contributed so much to the development of genetics discard information that had been accumulated and repeated for 50 years? Maybe, Mendelian genetics was a hoax and geneticists were stupid and dishonest. It seemed clear that the geneticists should answer the challenge. In December 1949 I attended in New York the annual meetings of the Genetics Society of America. After considerable discussion of the state of genetics in the Soviet Union, it was moved that a committee of eminent geneticists be elected *to speak for the society.* This committee would carefully study the evidence and make a statement for the rest of the geneticists of the society and the world. The motion was put to a vote. (How would you vote? Why?) The motion was defeated and for the following reasons!

1. To appoint a committee to speak for the society would be to do what the Russians had done. Each individual must maintain the right to speak for himself. The Royal Society of London, a most prestigious scientific body, has never appointed a committee to speak for the society. When asked to express an opinion, the problem is turned over to a committee that speaks for itself and not for the society. There might be some members of the Genetics Society of America who agree with Lysenko. There can be no official view to which all geneticists must bow.

2. A society cannot pass a decree involving the regulation of nature. "It is hereby moved that water consists of oxygen and hydrogen." "It is hereby declared as our official position that cells divide by mitosis." The Central Committee of the Communist Party decreed that the Lysenkoist view of genetics was true. How would you view a resolution by Congress that man arose from lower organisms by evolution? The official view of the Communist party was that genes do not exist!

3. The question is not whether the Russians were right or wrong scientifically. We all make mistakes. If the Russians are right, they will be corroborated; if wrong, they will be refuted. It is a matter to be settled in the laboratory. But even if the Russians were right scientifically, they would still be wrong because they had tried to accommodate a law of nature into a prevailing philosophical context. To the Russians it was a political, rather than a scientific, matter. It is politics, not science.

4. Why should we limit our science? If there is scientific justification of the Soviet views by genetics, then these will be accepted and will be incorporated into our present science of heredity. Science and research are open-ended, and any idea, however well-established, can be challenged. It is the spirit of the scientific community to maintain an environment of academic freedom. In too many instances, ideas that had been thought to be silly were later shown to be reasonable and valid.

These opinions expressed by the membership of the Genetics Society of America give an excellent example of the spirit and attitude of science in its search for the laws of nature. Lysenkoism has now been discarded in genetic laboratories in the Soviet Union. There has been a return to Mendelian genetics and active research in molecular biology. The acceptance of the concept that genes and chromosomes exist has not destroyed communism or the concept of dialectical materialism.

* * * *

Concerning a Statement of Beliefs

You can be sure that no scientific society will ever ask its members to sign a statement of beliefs or to inquire concerning church membership or race. As a member of the genetics society, I am not required to believe there are genes; as a member of the Society for the Study of Evolution, I am not required to accept the concept of evolution. And yet there are certain "scientific societies" that require such a pledge of faith as a condition of membership.

The Creation Research Society requires its members to subscribe to a set of beliefs. These include:

1. The Bible is the written word of God. It is historically and scientifically true. The accounts of origins in Genesis are fact.
2. All basic types of living things were made directly by God during the Creation Week.
3. The Great Flood was a worldwide event.
4. Adam and Eve were specially created and sinned in the Garden of Eden.
5. The concept of evolution is not valid.

It is well to have a code of beliefs to live by in order to make oneself a "better person" and to gain emotional security. This is all right if one wishes to join a religious group but is not all right if one wishes to join a scientific group. The spirit of science involves the right to challenge *all* concepts and ideas. This is a major difference between science and traditional religion.

There are certain aspects of Lysenkoism that are like the Creationist position. There is the appeal to authority (Marx, the Bible). There is accusation of insincerity and heresy for those who disagree. There is the objection to the concept of chance. To the Theist, chance lacks purpose and is no explanation of natural law. To the Communist, chance is not an explanation because it lacks direction (purpose?); a person who accepts such as an explanation is a "shame-faced materialist" and such a person might just as well accept God which also is no explanation.

The concept of the gene was not a philosophical-political matter outside of the Soviet Union. The concept of evolution is not a religious-political matter except for certain sects in the United States. Millions of people and many sects believe in God and still accept the concept of evolution.

It may be useful at this time to compare and contrast the genetics controversy in the Soviet Union with the evolution controversy in the United States and the Galileo controversy in Italy.

CHAPTER 8

BIOLOGICAL AND CHEMICAL HOMOLOGY, A STUDY OF SIMILARITY AND COMPLEXITY

IN THE discipline of *historical linguistics* comparisons are made among languages in order to determine their relationships. French, Spanish, Italian, and Portugese are classified as romance languages and all were derived from Latin. This relationship is shown by the similarities in words and sentence structure of these modern languages. It is recognized that languages change in time and place and that geneological language trees can be constructed. These geneological relationships can be shown even for languages whose common source is unknown. It is interesting to me that the principles of similarity and complexity used by the biologist and chemist are also used by the linguist.

> The *Sanscrit* language, whatever be its antiquity, is of a wonderful structure, more perfect than the *Greek,* more copious than the *Latin,* and more exquisitely refined than either, yet bearing to both of them a stronger affinity, both in the roots of verbs and in the forms of grammar, than could possibly have been produced by accident; so strong indeed, that no philologer could examine them all three, without believing them to have sprung from some common source, which, perhaps, no longer exists: there is a similar reason, though not quite so forcible, for supposing that both the *Gothick* and *Celtick,* though blended with a very different idiom, had the same origin with the *Sanscrit;* and the old *Persian* might be added to the same family, if this were the place for discussing any question concerning the antiquities of *Persia.*
>
> —*Sir William Jones (1799) Third Anniversary Discourse of the President of the Royal Asiatic Society, London Vol. 1, 19-28.*

BIOLOGICAL HOMOLOGY

When the biologist studies different kinds of organisms, he sees not only structural differences but also similarities. He recognizes also that the structures are complex and that the developmental processes that gave rise to these structures are even more complex. When he sees, therefore, that different kinds of organisms are similar in certain complex structures, he concludes that they have inherited similar patterns from common ancestry rather than that these patterns have arisen anew or by chance. This similarity of structure and developmental process is called "homology."

Some similarities may arise, however, not only by means of inheritance from common ancestry (homology) but also by means of adaptation to a common environmental experience (analogy). The body form of the dolphin, shark, and the penguin are streamlined in response to the necessity to move in an aquatic environment. This similarity of the body form for dolphins, sharks, and penguins is an example of convergent evolution. It did not arise from a common ancestor that had fish-like form. Although superficially the flippers of dolphins are like the fins of sharks, and the wings of birds and insects have similar form and function, they are very different in their structure. Birds and insects do not share a close common ancestor from whom they inherited wings.

The biologist uses homologies in order to work out a record of ancestral relationships (phylogeny). In the study of the vertebrates there are many homologies for systems from fish to man. There is not only similarity for each complex structure but also for an array of such structures from bones, to kidneys, to the brain, to the arteries.

The forelimbs of vertebrates show a similarity of structure for animals of widely different forms and ways of life. The wing of a bird is used for flight; the flipper of a whale is used for movement through water; the forelimb of the horse is used for running; yet all of these forelimbs are built on the same body plan. There is present in each a common array of bones: the humerus, radius, ulna, carpels, metacarpels, and phalanges. In each organism, these bones have undergone modification to adapt the limb to a particular environment. The best explanation at present for this remarkable similarity of these bones in birds, whales and horses is that these creatures share a common ancestry.

One has never observed amphibians giving rise to reptiles nor reptiles to birds and mammals. Given the marked differences in their outward appearance, the hypothesis of evolution seems incredible. How much more incredible, however, is the process that we can observe, the formation of a complex chick from an egg. The range of differences between the fertilized egg and the newly hatched chick is much greater than between the salamander and the lizard or the chimpanzee and man. The differences between a caterpillar and a moth or between a tadpole and a frog are so great that without a knowledge of their life cycles we would not consider them to be the same organism. Development is a process that is so amazing that if we did not see it everyday we would not believe it.

A considerable effort has been put forth by biologists in the analysis of development in different species. In these studies one looks for homologies in changing structures and processes. The developmental patterns, just like the adult structures, have been inherited in different embryos from a common ancestry, and because of this one finds that patterns found in man are like those found in fish. The analysis of development (ontogeny) gives insight into ancestral history (phylogeny). Mammalian embryos recapitulate (repeat) many of the events of fish embryos. The embryologist, Karl von Baer, 1828, recognized that the early embryos of a higher species resembled the early embryos of lower taxonomic species.

One can observe that all young embryos are similar in appearance, but as they go through development they become different. Comparison of vertebrates from fish to man shows that their early stages resemble one another to such a degree that it is difficult to distinguish among them. All vertebrate embryos show the phylum Chordata characteristics, consisting of a dorsal tubular nerve cord, notochord, and pharyngeal gill slits. As development proceeds one can recognize fish from chick and calf from man embryos.

There are many examples where homology in the developmental events repeat in different species and these give us confidence in our concept of evolution. Some of these examples are discussed below. The explanatory diagrams can be obtained from standard biology texts.

The Vertebrate Kidney

Among the vertebrates there are three types of kidney, the pronephros, the mesonephros, and the metanephros. These three types of kidney appear sequentially during the development of a mammal, but only one of them becomes the functional adult kidney. In the development of the mammalian kidney, the pronephros is the first to appear; then the mesonephros is formed and the pronephros degenerates; then the metanephros appears and the mesonephros degenerates. In different classes of vertebrates, different types of kidney become the adult kidney. When one compares kidney development in different classes, one finds that the more primitive vertebrates possess a kidney that occurs earlier in development, while the later occurring vertebrates possess a kidney that occurs later in development. Among the jawless fishes, such as the lamprey, the pronephros is the larval kidney and the mesonephros is the adult kidney. Among reptiles, birds, and mammals the metanephros is the last kidney to appear and is the functional adult kidney. Reptiles, birds, and mammals develop pro-, meso-, and metanephric type kidneys; jawed fish and amphibians develop the pro- and mesonephric types. Thus, the higher vertebrate embryos pass through the developmental events that are experienced by more primitive vertebrate embryos.

The Mammalian Jaw and Middle Ear

Mammals have a middle ear consisting of three bones, the malleus, incus, and stapes, and a lower jaw made up of a large single bone, the dentary. Reptiles have a single bone, the stapes in the middle ear and a lower jaw made up of the dentary bone and several post-dentary bones. The jaw joint of mammals is formed between the temporal bone of the skull and the dentary bone of the lower jaw. In reptiles, the jaw joints lie between the last post-dentary bone, the articular, and a skull bone, and quadrate. During the development of the mammalian jaw there is a stage that is comparable to the reptilian structure. Observations of development on suckling marsupial embryos have shown that the jaw joint is made up of the articular and quadrate bones. This is a jaw joint like that of reptiles. During later development the articular and quadrate bones are incorporated into the middle ear as the malleus and incus. The dentary bone increases in size to form the entire lower jaw which joins with the temporal bone.

That the ontogenetic processes of the historically younger vertebrate groups build upon the patterns of the older vertebrate groups is a recognizable phenomenon. This holds not only for kidneys, jaws, and ear bones but also for all other structures.

Developmental biologists have approached the problems of ontogeny and phylogeny not only by the observation of developmental events in the different groups but also by experimentation. These experiments demonstrate that developmental processes have the capacity to bring about structural changes of the magnitude of class differences among vertebrates.

The Body Covering of Reptiles, Birds, and Mammals

Scales, feathers, and fur are comparable structures and develop in similar ways. It is thought that feathers and fur arose during evolution from reptilian scales. Scales, feathers, and fur are composed of similar proteins, the keratins. Occasionally, there is a transformation of the tips of scales into feathers on the legs of birds.

The skin is made up of two layers. The superficial layer, the epidermis, develops from the ectoderm. The underlying layer, the dermis, develops from the mesoderm. Scales, feathers, and fur develop from the epidermis but depend on signals received by the dermis. When epidermis from a region of the body that produces feathers (dorsal region) interacts with dermis from the dorsal region, feathers are produced. When epidermis from the legs (tarsometatarsal region) interacts with tarsometatarsal dermis, scales are produced. However, dorsal epidermis grafted onto tarsometatarsal dermis produces scales and tarsometatarsal epidermis combined with dorsal dermis produces feathers. Thus,

1. dorsal epidermis and tarsometatarsal dermis—large scales;
2. tarsometatarsal epidermis and dorsal dermis—feathers.

The kind of epidermal structure produced depends upon a signal that is sent from the underlying dermis.

Experiments have been done on reptiles, birds, and mammals in which the epidermis and dermis have been combined.

1. Lizard epidermis and chick tarsometatarsal dermis produces large scales like chick scales.
2. Lizard epidermis and chick dorsal dermis produces arrested scales arranged in a typical hexagonal feather pattern.
3. Lizard epidermis and mouse dermis produces scales arranged in typical mouse hair patterns.

These experiments demonstrate that the production of highly differentiated structures, such as feathers and scales, depend upon subtle signals. There is no fundamental difference in the tissue that forms a scale or a feather; its production depends upon a signal produced during the differentiation process. Such experiments demonstrate that a reptile

TABLE II

INTERACTION BETWEEN
A FEATHER-PRODUCING REGION, DORSAL (D), AND
A SCALE–PRODUCING REGION, TARSOMETATARSAL (T)

Epidermis	D	T	D	T
Dermis	D	T	T	D
Structure Produced	Feathers	Scales	Scales	Feathers

Epidermis	Lizard	Lizard	Lizard
Dermis	Chick T	Chick D	Mouse
Structure Produced	Large scales like chick	Arrested scales in feather pattern	Scales arranged mouse hair pattern

with feathers or with fur could be produced and suggest a similarity (homology) between reptiles and animals with feathers.

The Teeth of Birds and Mammals

The development of teeth depends upon the interaction of oral epithelium and oral mesenchyme. Birds do not have teeth, and the interaction of oral epithelium and oral mesenchyme produces a beak. Chick oral epithelium and duck oral mesenchyme produces a duck beak. Duck epithelium and chick mesenchyme produces a chick beak. Chick oral epithelium was combined with mouse oral mesenchyme and these tissues were cultured together. They produced tooth enamel substances and in some cases complete teeth. Thus, the epithelial layer of birds still has the ability to form teeth, but a change has occurred within the mesenchyme so that it no longer can signal the formation of teeth.

Vestigial Structures

Structures and organs that are non-functional and rudimentary appear in many organisms. Some of these examples are presented below. A rudimentary pelvic girdle is embedded in the body of whales. It is not attached to the vertebra and is of no use for locomotion. This is taken as evidence that the whale evolved from a four-legged ancestor. Some snakes such as the python and the boa show rudimentary hind limbs.

Birds produce large eggs which contain structures that support the maintenance of the developing embryos. The embryo is enclosed by an amnion and protected by amniotic fluid. Nutrition is obtained by means of food (yolk) contained within a large yolk sac. Waste products are collected in a large sac-like allantois. The entire egg is surrounded by the chorion which contains many blood vessels in order to allow for the exchange of oxygen and carbon dioxide. In most mammals, including man, the amnion, yolk sac, allantois, and chorion are present. The yolk sac and allantois, however, are not functional and are rudimentary.

Organisms sometimes show structures that resemble those of their ancestors. These "throwbacks" (atavisms) may involve large changes, because a small change in an embryo may be amplified as development proceeds. Stephen Gould has reported that Caesar's horse had hoofs that were cleft-like toes. The paleontologist O. C. Marsh (1982) studied horses with extra toes, where the side splints had developed into functional, or nearly functional, hoofed toes. The fly *Drosophila* is a member of the order Diptera, two-winged flies, which is characterized as having one pair of large wings and one pair of small wings (halteres). Other orders of insects possess two pair of wings. A gene in *Drosophila* has been discovered that is capable of changing development so that four wings of equal size appear. A group of genes (homoeotic mutants) are capable of producing legs in place of antennae. The developing embryo is capable of a wide range of events to produce marked structural changes. The change from the forearm of a land vertebrate to the flipper of a whale is remarkable. The change from a fly with two wings to four wings or from antennae on the head to legs on the head is also incredible, but here it is.

CHEMICAL HOMOLOGY

Studies by biochemists over the past 50 years have established that the basic metabolic processes of all living organisms are very similar. This comparison of chemical structures and pathways covers the range from viruses and bacteria, to plants to vertebrates. It needs to be emphasized that these large molecules are complex and that their interactions in metabolism are complex.

1. All cells utilize polyphosphates, particularly adenosine triphosphate, for energy transfer. These polyphosphates are made in photosynthesis, or the oxidation of stored food.

2. Cells make, store, and degrade fats, carbohydrates, and proteins in similar ways.
3. Metabolic reactions are catalyzed by proteins. Homologous proteins such as the respiratory molecule cytochrome c can be functional in organisms from yeast to rat.
4. Proteins are made by a coding process in which the sequence of codons in the DNA determines the sequence of amino acids in the protein.

This genetic code is universal, in that the twenty basic amino acids are made by the same codons. This universality is highly improbable by chance alone and points, therefore, to an explanation based on the common ancestry of present-day organisms and probably one grand phylogeny. Although the number of chemical steps between a gene and a bone are many, the number from a gene to a protein are relatively few. When one studies changes in amino acids in proteins, one can therefore interpret these amino acid changes in terms of codon changes. An amino acid substitution in a protein results from a nucleotide substitution in a DNA molecule. When one compares the same kind of protein (protein homology) in different species, one is comparing also their genes and their gene mutations.

Proteins, although chemicals, are complex structures and each amino acid position in each protein may be considered as a variable trait. A protein that is 100 amino acids long could consist of 20^{100} variations, since there are 20 common amino acids. These proteins are subject to natural selection, in that the more efficient variations are preserved and inherited. There is an enormous amount of information in the proteins. The biochemist has as the goal to work out the phylogenetic tree and to see how this lineage elucidated by chemistry compares to that worked out by biology.

Cytochrome c is a respiratory pigment that is found throughout the microbial, plant, and animal kingdom. Substitution of cytochrome c from different species as shown by test tube reactions shows the same function and thus the cytochrome c in all organisms is homologous. The amino acid sequence has now been determined in 40 species (e.g. yeasts, moths, tuna, chicken, dog, man) and the similarity of cytochrome c as determined by the position of the amino acids follows the phylogenetic relationships worked out by biologists. The cytochrome c in man is identical with the chimpanzee and differs from the horse by 11, dogfish by 24, bread mold by 44 positions.

Similarity and Ancestry

In classification, organisms are placed into groups depending upon their similarities and differences. In this system of hierarchies there are groups within groups. A goldfish is different from a kangaroo, in that it has fins, gills, and a two-chambered heart. It is similar to a kangaroo, in that it has vertebrae, a brain case, and a dorsal tubular nerve cord. Both the goldfish and the kangaroo are members of the phylum Chordata and of the subphylum Vertebrata, but they belong to different classes (Osteichthyes and Mammalia). According to evolutionary theory, these similarities between the goldfish and the kangaroo tell us that there was once a common ancestor that showed these phylum characteristics.

A system of classification is not designed only for the identification and cataloging of organisms. It is also meant to show genetic relationships. All vertebrates are related by common ancestry. All birds are related by common ancestry. The common ancestry of vertebrates existed further in the past than the common ancestor of birds. It needs to be emphasized that present-day species did not descend from present-day species. Present-day mammals did not descend from present-day lizards. Both mammals and lizards have evolved from a common ancestor that was different from either of them.

Whenever one finds different species that are similar in certain complex traits, one looks for a common ancestor that shows these traits. Cell nuclei are found for a wide range of organisms (vertebrates, invertebrates, protozoa, higher plants, many algae, yeasts), while feathers are found only for birds. The existence of feathers tells the biologist that birds have descended from a feathered creature. The existence of nuclei means, in evolutionary theory, that all organisms with nuclei were derived from a common ancestor that possessed nuclei. Nuclei are more basic than feathers, because they encompass a wider range of organisms and point to a common ancestor in the more distant past. Pharyngeal gill slits are more basic structures than hair. Hair is a characteristic of the class Mammalia, while pharyngeal gill slits are characteristic of the entire phylum Chordata of which Mammalia is a part. The common ancestor of chordates is older than the common ancestor of mammals. The existence of ATP as a complex energy-generating mechanism for organisms is older than the existence of nuclei. Some groups (prokaryotes) do not possess nuclei, but all organisms utilize ATP. All organisms have the same genetic code. This is taken to indicate common ancestry for all organisms.

Thus, the similarities of complex structures and chemical processes are used to determine ancestral relationships. The more basic traits appear earlier in development and also have appeared earlier in the evolutionary process. It is interesting and instructive to inquire not only in what way elephants and oak trees are different but also to ask how they are similar. There are many obvious differences between a frog and an ameba. In what ways are the frog and the ameba alike?

Some Detective Work

Three different houses were robbed during the month of December. In each house fingerprints were found on the window panes of opened windows and the fingerprint patterns were found to be identical. A suspect is found whose fingerprints are identical with those found on the window panes. The police arrest him and charge him with the robberies. The suspect claims that he is innocent, that he was out of town on the days of the robberies, that no one saw him commit the robberies, and that the identical fingerprint patterns was a coincidence. As the prosecuting attorney, how do you build your case against the suspect. As the defending attorney, how do you build your case in favor of the suspect.

Scholars involved in the study of mythology find similar myths among different peoples and use this similarity to show how one myth is derived from another. Do these principles of similarity and complexity apply to other fields such as music, art, and law? Is this a valid way to elucidate relationships and origins?

CHAPTER 9

THE FOSSIL RECORD AND THE AGE OF THE EARTH

ABOUT 200 years ago, everyone believed that the earth was about 6,000 years old. This estimate was based upon counting back generations to the creation of the earth and Adam as revealed in the Bible. Since then, other methods for determining the age of the earth have been discovered. The general trend in these estimations has been toward an older and older earth. Geologists, by their methods, could make estimates in the range of hundreds of thousands of years; physicists by measuring the rate of cooling of the earth had estimates from 10-20 million years. The discovery of radioactivity at the end of the nineteenth century led to estimates in the billions of years. The earth and solar system is only a small part of the universe and is estimated by astronomers to be younger than the universe. With the discovery of new scientific methods, these estimates may need to be revised, but these revisions almost certainly will not return us to a scientific agreement that the earth is only 6,000 years old.

As we push back and back in time we will always find an earlier time shrouded in mystery. But this is the frontier that one faces in scientific discovery, not only in the age of the earth, but also the size of the universe, the structure of the atom. To believe that we will never know everything is not contrary to the knowledge that we have learned much, nor to the spirit that we should continue to try to discover. When we look over our scientific accomplishments over the past 400 years, we must be impressed by our discoveries and inventions. Unanswerable questions of one generation are solved in another generation. It is difficult to think of any specific unanswerable questions of today that may not be answered in the future.

A geological time scale has been established and used to record the history of the earth. Absolute dating methods involving the degeneration of radioisotopes places the age of the earth at 4.5 billion years. Geologists have divided this history into eras and these eras in turn have been divided into periods and epochs. The time durations for each time interval and the fossil creatures found for that time have been determined.

The earth's crust is laid down in layers. Geologists established that the lower layers are the oldest and the upper layers are the youngest. This dramatic stratification can be seen in the Grand Canyon. The lowest layer consists of basalt, a rock arising from volcanic activity. It is 2 billion years old. The top layer, on which the campground is located is 200 million years old. At one time, layers of rock younger than 200 million years were present in the Grand Canyon on top of the present layer, but these were eroded away by the action of water, ice, and wind. We know this because in the larger rock formation of which the Grand Canyon is a part, these younger and higher layers are found. In the younger rocks of Utah, dinosaur fossils occur, but there are no dinosaur bones in the older rocks of the Grand Canyon. One can obtain estimates of the age of each layer relative to the other layers by measuring the thickness of each layer and the rate of sedimentation of sand and lime. Measures of time are also obtained from the rate at which water cuts through the rock. The Grand Canyon has been made by the Colorado River as it moves rapidly from a 12,000-foot altitude in the Rocky Mountains to sea level in the Gulf of California. In this torrent of water, millions of tons of rock and silt are swept down the river. Geologists have estimated that it took the Colorado River about 6 million years to cut through the rock to its present level. It is a glorious experience to descend to the bottom of the Canyon and to cover one mile in altitude and 2 billion years in six hours of hiking. The returning ascent, however, is not quite so glorious, particularly in hot weather.

The layers of sedimentary rock in the Grand Canyon were laid down by the sea. Within the different layers, marine fossils are found. Sand, for example, settles in water and eventually, due to packing, becomes hardened. The pressure of one layer upon the other changes the soft sand into hand sandstone. The Grand Canyon has alternated between times when it was above and below sea level. Visible are the remains of organisms that could have lived only in water.

People who study ancient life (paleontologists) are interested in rock formations because they contain fossils. Fossils are remnants of organisms or parts of organisms or casts of organisms that lived long ago. It was established about 200 years ago that specific layers of rock contained specific kinds of fossils. These fossils are not repeated in other layers. Thus one can identify the age of fossils by the age of the layer of rock. Certain fossils are of worldwide distribution and abundant. These fossils are known as index fossils, because they can be used to date layers of rock in different parts of the world. The index fossils are marine in origin and therefore can have a worldwide distribution. During the Cambrian period, trilobites are found; during the Silurian period, graptolites occur; during the Cenozoic era, foraminifera; and so forth.

At the beginning of the Paleozoic era, the Cambrian period, 600 million years ago, many types of complex organisms were present in abundance. The dominant groups at the time were the marine invertebrates. From this time one can recognize a succession of kinds of organisms with a succession of layers of rock. The earliest fish appeared in Ordivician (480 million years ago) and flourished during Devonian (400 million years ago). The vertebrates invaded land and this led to a succession of amphibians, reptiles, birds, and mammals. Man appeared during the Pleistocene epoch, 2 million years ago. The evolutionary scheme would receive a considerable jolt if amphibians were found in Precambrian deposits, if mammalian fossils were found in Devonian deposits, or if human fossils were found in Mesozoic deposits with dinosaurs. Such new observations would need to be published in refereed scientific journals, and their acceptance by the scientific community would depend upon corroboration by other independent research laboratories. If legitimate human footprints were found associated with dinosaur footprints, these photos would appear on the cover of *Science* regardless of their impact on the concept of evolution. It would be one of the most important discoveries of the twentieth century. This observation would stimulate a great amount of research and change our scientific views concerning the history of living things.

When fossils are collected from successive layers of earth, one looks for progressive changes in structure. Among vertebrates, these studies usually involve bones and teeth because these structures do not decompose as readily as softer tissues. One also looks for transitional forms between groups of organisms. There is a limitation to fossil study, since the fossil record is not sufficiently complete to work out complete phylogenies. Most organisms are not fossilized after death, they decompose. However, some groups of fossils occur in abundance and small changes can be recognized from one structure to another. Although the fossil evi-

TABLE III

THE GEOLOGICAL TIME SCALE
(After Volpe)

Era	Period	Epoch	Millions of Years Ago	Important Events
Cenozoic	Quarternary	Recent	0.01	Modern man
		Pleistocene	2	Early man
	Tertiary	Pliocene	6	Radiation of apes
		Miocene	25	Grazing mammals in abundance
		Oligocene	38	Flowering plants dominant
		Eocene	55	Radiation of mammals
		Paleocene	65	First placental mammals
Mesozoic	Cretaceus		135	First flowering plants
	Jurassic		185	Reptiles dominant, first birds and first mammals
	Triassic		225	First dinosaurs
Paleozoic	Permian		275	Radiation of reptiles
	Carboniterous		350	Swamp trees, amphibians prominent
	Devonian		400	Age of fishes
	Silurian		430	Invasion of land by plants
	Ordivician		500	Earliest jawless fishes
	Cambrian		550	Marine invertebrates, algae
	Ediacaran		670	Soft-bodied invertebrates
Precambrian			2500	Evolution of cellular structures and processes
Hadean			4600	Origin of the Earth, formation of a crustal layer, chemical evolution

dence is not sufficiently complete to determine precise phylogenies, it is adequate to show the phenomenon of evolution. New fossil beds are continually being discovered and studied. Paleontology is an active field.

The existence of fossils and the knowledge that fossils were remnants of extinct organisms was known before the time of Darwin. It was taken to mean, however, that all life had been destroyed from time to time by catastrophies and that new life was created from non-living material of from a few survivors. But too many catastrophies were needed to explain all of the extinctions, and not all groups of organisms living at a particular time died out at the same time. Only certain groups became extinct at a particular time. When the dinosaurs died out, mammals, birds, and other reptiles survived.

George Gaylord Simpson (1949), upon examination of the fossil record, has made certain generalizations concerning the history of organisms in geologic time.

1. Different groups of organisms appeared at different times. Water forms appeared before land forms; invertebrates before vertebrates; reptiles before mammals.

2. There has been a succession of successful types. These successful types have flourished because of certain innovations that gave them an advantage over their competitors or permitted them to exploit a new environment free from competition. The innovation of lungs enabled fish and amphibians to exploit the land. The innovation of skin that prevented water loss and of an amniote egg freed the reptiles from water. The innovation of a placenta and maternal care enable more mammals to survive and multiply. The brain in man enabled him to become the dominant organism today.

3. There has been an increase in the complexity of organisms. Useful innovations have tended to become maintained and improved.

4. The evolutionary rates have not been the same for all species nor constant in time. Some species living today have not changed for millions of years. The horseshoe crab is of the same appearance as it was 200 million years ago. The opossum has not changed for 75 million years. Other species have changed rapidly when they exploit a new environment. Such adaptive radiations ("evolutionary explosions") occurred for the marsupials in Australia where they adapted to many different unexploited environmental niches. They occurred for the reptiles and for the mammals. Evolutionary change has been rapid in mammals and slow in birds and amphibians.

5. Many species leave no descendents at all. They become extinct. Only a few groups have multiplied greatly and the inhabitants of any period are therefore descended from a few groups. Dying-out groups are

replaced by surviving ones, reptiles to other kinds of reptiles or to entirely new organisms that have no direct descendents living today. The vast majority of species have become extinct.

6. Evolution depends on opportunity. It depends upon the ability of the species to exploit its environment. The species can only react to its immediate environment. It cannot anticipate what the environment will be like 1,000 years hence. There is no built-in vital force (entelechy) that directs evolution to specific ends.

The Age of the Earth-Science or Faith

Different religious groups have each determined the age of the earth. According to Orthodox Hindus who follow their sacred books, the Vedas, the earth was created not earlier than 1500 B.C. To Christian Fundamentalists, the earth is 6,000 to 10,000 years old. What is the basis for these estimates of the age of the earth? How does one judge which of these estimates is correct? How does one judge between these "truths?" The matter of Creationism and the age of the Earth has been addressed by Philip Abelson, editor of the magazine *Science,* in 1982. A summary of his paper follows.

An enormous amount of data collected by astronomers, physicists, and geologists dealing with the age of the earth are in agreement that the earth and universe is much more than 10,000 years old. The age is estimated to be about 4.5 billion years. Ancient rocks have been dated by radiometric methods for 5 different radioisotopes. One of these methods measures the rate at which uranium-238 decays to lead-206 and uranium-235 decays to lead-207. By this method, the ages of certain rocks 2,600 million years old have been determined with a probable error of 1 million years. Geophysicists who study the rate at which continents are moving apart have determined that it took 250 million years for South America, Africa, Australia and the other large land masses to have moved to their present positions from a time when these land masses were contiguous. Geologists have found a large number of formations in which the total number of annual deposits due to glaciers far exceeds 10,000. Geochemists have found that the process by which petroleum is formed from organic matter requires more than 1 million years. Thousands of scientists have produced evidence relevant to the age of the earth. These data have been independently obtained by a wide variety of methods. If the age of the earth were 10,000 years, it would be accepted by all scientists. There is no substantial body of experimental data to back the creationist view. The truth is not on their side and their activities can only harm their cause.

In a recent issue of *Science* (Kerr, 1983), there is a report about the measuring of time in marine carbonate sediments. The techniques that they used enabled them to go as far back as 800,000 years with an accuracy of 3,000 to 5,000 years (800,000 ± 5,000 years). This dating technique relies on the record of oxygen isotopes and the periodic orbital variations of the earth.

On what basis is the citizen to make a judgment concerning the age of the earth? Is he being asked to make a choice between two authorities? Most citizens do not understand the theory and methods by means of which the age of the earth was determined. This information is not a secret. It is available to anyone who wishes to do the hard work necessary in order to understand. No one can understand the theory and methods of all scientific phenomena (photosynthesis, atomic theory, DNA) and yet he needs to be aware of these phenomena to shape his world view. He needs to ask some questions.

1. Does the proponent of a point of view have an axe to grind? Does he relate his opinion to morality and religion?
2. Is there a substantial quantity of theory and data?
3. Do the data have quality and are they subjected to critical analysis? Is there a spirit of critical analysis in the enterprise? Have the estimates concerning the age of the earth been arrived at by different independent methods? Have the experiments been repeated by different laboratories and are they in agreement? Can the explanations fit into the theory of the science? If not, can this be explained?
4. What is the "track record"? How many times in the past has a religious explanation been used to discard a scientific explanation? How many times is an infallable source permitted to be wrong before it loses credibility?
5. Are opinions concerning the age of the earth subjected to threats or injury? In 1834, Thomas Cooper, a chemist, was fired from his position as President of the College of South Carolina for publically stating that the earth was much more than 6,000 years old. Archbishop Ussher of Ireland stated in 1654, "The world was created 9 A.M., October 26, 4004 B.C. Anyone who disputes this is guilty of heresy."
6. Are derogatory phrases used as arguments? In a scientific discussion, how does one respond to terms such as *hoax, capitalistic geneticists, atheistic scientists, secular humanists, heretics?*

Scientific estimates of the age of the earth have been criticized for about 200 years for religious reasons. According to the Eckanckar religion, man arrived on earth from outer space about 200 million years ago. Is the age of the earth simply a religious matter or may one do research in this field of human endeavor? Should we write different "scientific books" for Hindus, Christians, Buddists, Moslems?

Christopher Morley, a twentieth century english author summed up the matter, "Where it is a duty to worship the sun it is pretty sure to be a crime to examine the laws of heat."

Changes in Time and Space — Horse Evolution

An interpretation of evolutionary events by the study of fossils rests on the remnants of plants and animals that have been collected in different parts of the world and at different geological times. We are dealing with changes that are too slow to be observed in a human lifetime; changes that have occurred over millions of years. Analysis involves the study of sequential changes in fossils with time. This approach is shown on the evolution of the horse. This history of horse evolution covers a period of 60 million years. It is considered by paleontologists to be one of the best documented examples of evolution. This history of the horse begins in Paleocene. In its evolution a large number of species were formed and later became extinct. Today, there exists five species of horse: the donkey, the kiang of Mongolia, the mountain zebra, the plains zebra, and the mongolian horse from which our domestic horses have been derived. The horse originated in North America but spread by means of the Bering Strait to Asia and then to Europe. The North American horses later died out and all present-day horses are the progeny of domesticated horses brought in by the Spaniards. Present-day wild horses in the United States have flourished. Why the horses at one time died out is not known.

Present-day horses arose from a common ancestor, the dawn horse (Eohippus = Hyrocotherium). The differences between the Hyrocotherium that lived 60 million years ago and the present-day horse are so great that when these fossils were first found, they were not placed into the horse phylogeny. Gradual changes in structure were observed in fossils, however, and a continuous line could be established back to Hyrocotherium. It was not until these connections through intermediate fossils were made that it could be recognized that Hyrocotherium (now = Eohippus) was the common ancestor to present-day horses. The appropriate diagram can be found in biology textbooks.

The earliest horse was only about 12 inches in height. It had 44 teeth with low crowns. From these teeth one can determine that this animal fed on leaves. Each forelimb had 4 toes and 1 splint, while each hind limb had 3 toes and 2 splints. It lived in the forest.

An environmental change from forests to plains occurred where these early horses were living. The food now consisted of grasses grown in sandy soil. In the plains where horses could now easily be seen, there was an advantage to be able to run swiftly in order to escape from predators. During this period there has been an increase in size. A longer and more mobile neck developed; limbs became longer; the third toe became enlarged to form a functional hoof, while the other toes were lost. There was a change in diet from soft leaves to sandy grasses. In response to this diet, teeth with higher and cement-covered crowns have emerged. The molars developed into grinding structures; the teeth became less in number.

What is one to make of this? To the biologist this is a prime example of evolutionary change. Many fossils of horses are found from one stratum to the next and the migration patterns can be determined. Evolution can be seen as a continuous process and it is impossible to perceive where one line ends and the other begins. To the creationist, each group of horse was created separately at each separate time. Or, it has been argued that this is not an example of evolution but rather is variation within a kind; after all, they are still horses and the orderly sequence of structural change with time is merely coincidental. One needs to point out here that the orderly sequence of change has occurred for other groups of organisms. If the process of evolution did not occur, one would not expect to see an orderly sequence of change with time and strata. We did not know that "they are still horses" until we put the sequence together.

Go back to "picture frames in onions" and compare the analysis of mitosis by "frozen cells" to horse evolution by "frozen fossils." Should we consider the sequencing in time of individual static fossils to be a valid approach to the study of change within time and place? Can one record changes that have occurred over millions of years when such changes are imperceptable during one's lifetime? How is one to analyze historic events that were unseen and not repeatable?

If present-day organisms have descended from other organisms that were not like them, what would one expect to find in the fossil record? This question was addressed by John A. Moore (1983) who discusses the following criteria:

1. One must be able to study organisms that lived in the past. This is done by the analysis of fossils and the layers of rock that make up the earth.
2. Species that lived in the past must be different than those that live today.
3. The simplest organisms should be found in the oldest strata and the more complex organisms in the more recent strata.
4. It should be possible to demonstrate the slow change from one species to another.
5. It should be possible to find transitional forms between phyla and classes.
6. The earth must be very old.
7. Homologies in structure and development must exist for different classes of organisms.

Mantracks in Texas

In the magazine, *Natural History,* Vol. 95, August, 1986, there is an article *"Blunder in Their Footsteps"* by Godfrey and Cole which discusses the finding of dinosaur tracks and mantracks in Glen Rose, Texas and the claim that dinosaurs and humans coexisted. This can be the start of a class project to focus on evidence that has been studied by both the creationist and the evolutionist. Its study will be a rewarding experience.

If it is true that men and dinosaurs were contemporaries, this would have to be reconciled with the many observations that different kinds of organisms are found in different layers of the earth's rock.

CHAPTER 10

THE EVOLUTION OF THE VERTEBRATES

THE VERTEBRATES were the last group of chordates to appear on earth. These different classes of vertebrates have first appeared at different times. The general scheme determined from geological and fossil evidence is that fishes were ancestors to amphibians; amphibians to reptiles; reptiles to birds and to mammals. In this scheme of evolution, one can recognize the occurrence of (1) innovation in structure, (2) adaptive radiation involving diversification of structure in the exploiting of new environments, and (3) transitional forms between classes and orders as represented by fossils and by present-day organisms.

In order to fully appreciate vertebrate evolution it is necessary to study the bones and teeth of both fossils and of present-day animals. Paleontology involves a study of both geology and comparative anatomy. A trained paleontologist can recognize differences between the bones and teeth of representatives of the different vertebrate classes. Recently, a group of paleontologists discovered a single fossil jawbone in the Antarctic. They were able to recognize that this was a marsupial bone and to determine that it is 40 million years old. Certain fossils can be prepared for the study of structural details under the microscope. Sometimes entire animals such as the mammoths preserved in glacier ice are found; occasionally complete skeletons are found, but most often bones or parts of bones are collected. Within museums of natural history throughout the world, the skeletons of dinosaurs and of other vertebrates have been assembled from many single bones. To obtain a fuller understanding of vertebrate evolution, one needs to turn to a text such as *Evolution of the Vertebrates, a History of Backboned Animals with Time* by Edwin H. Colbert (1980). If one wishes to put forth the effort to learn the bones and teeth, one can go back in time with the paleontologist and discover events that have never been seen by man. I have relied upon this book for this chapter. Appropriate diagrams can be found in this book and in other biology texts.

Although most of the remains of animals have been destroyed by scavengers and by degeneration with time, there are many excellent fossil beds where an abundance of many different kinds of well-preserved remains are found. The Solnhofen Limestones in Bavaria have produced excellently preserved fossils of fish, horseshoe crabs, lobsters, dinosaurs, and Archeopterix. Fine details can be seen in these fossils. Rich deposits of dinosaurs are found in Colorado and Utah. At the Dinosaur National Monument, within a single layer of stone, fourteen different species of dinosaur have been found. In the limestone caves near Adelaide, Australia, a large number of marsupial fossils were discovered in 1969. Rich fossil beds have been located in Greenland, Antarctica, South Africa, India, and the United States. The search for more fossil beds continues. With more explorations into regions of the earth that are now accessible to man, more groups of vertebrate fossils will be discovered and analyzed. The study of fossils is today a very active field of investigation. New discoveries relating to phylogeny are constantly being made.

The study of vertebrate evolution involves the analysis of transitional groups showing structures that appear in two different classes. There are fossils, for example, that show both reptilian and avian characteristics. There are fossils that show both fish and amphibian traits. As we shall see, fossils that show intermediate characteristics between classes have been found for all of the vertebrate classes. In addition, certain living species such as the coelacanth and the duck-billed platypus each show characteristics of two different classes. It is not true that there are no missing links. Vertebrate evolution also involves the appearance of enormous variety within each class. There are today about 40,000 species of bony fishes that have evolved from a bony fish ancestor. There are about 4,000 species of mammals that differ markedly, from elephants to whales, to bats, to lions, to anteaters, to man. All of these different mammals evolved from common ancestry. In some instances, remarkable new innovations arise such as the development of jaws or limbs or a land egg or intelligence and these give to the bearer a distinct advantage in survival. In other instances, structural and behavioral changes are brought about in response to new environmental challenges and the exploitation of new niches.

Fishes to Amphibians

The two major classes of fish today are the cartilaginous fish (sharks and rays) and the bony fish. It is the bony fish group that gave rise to the amphibians.

There are two groups of bony fishes: (1) the ray fins in which skeletal support for each fin is by means of slender bony rods that radiate out from the proximal part of the fin; and (2) the lobe fins in which each fin is fleshy and muscular and supported internally by a smaller number of larger bones. Among the lobe-finned fishes there was one group, the Crossopterygii, that seemed to be of evolutionary significance, because it shows a combination of fish and amphibian characteristics.

1. The Crossopterygii have paired fins in which a single proximal bone (humerus) joins with the pectoral girdle and another proximal bone (femur) that joins with the pelvic girdle. Joined distally to the humerus are the radius and ulna, and to the femur are the tibia and the fibula. This basic bone pattern seen in Crossopterygiian fossils is found in all of the other classes of vertebrates, fossils and living! The most distal limb bones of the Crossopterygii, however, are not clearly comparable to fingers and toes. It had been thought that all of the Crossopterygii had died out and existed only as fossils. A number of such fish named coelacanths have been caught off the coast of Africa. This "living fossil" looked like the paleontologists had predicted based on their study of fossils. An excellently preserved coelacanth, about three feet in length, is housed in the Harvard Museum of Comparative Zoology.

2. The bony pattern of the skull is much like that of the early amphibians and different from the skull of the ray-finned fishes. The two large parietal bones of the skull can be homologized with the amphibian parietal bones.

3. Crossopterygian fishes had teeth in the margins of the jaws for grasping prey. When these teeth are examined under the microscope, it can be seen that the enamel is highly folded. This kind of tooth, the labyrinthodont tooth, is known only in the Crossopterygii and the primitive amphibians.

4. There was a well-developed nasal passage from the outer nostrils to the mouth or pharynx such as is found in land vertebrates.

5. A pair of lungs as outpocketings of the alimentary tract were present.

6. The vertebrae are like those of the primitive amphibians.

From the foregoing it can be concluded that the Crossopterygii are clearly fish that possess certain amphibian features. They are, therefore, considered to be a transitional group. A group of fish, Eustenopteron, are considered by paleontologists to be on the direct line towards the early amphibians. It is this group that, during late Devonian, came out on land to start the events leading to transformation into amphibians. Fossils of primitive amphibians have been found in the Devonian sediments of Greenland. These ancient amphibians, Ichthyostegids, show both amphibian and Crossopterygian characteristics. The skull of the Ichthyostegid has a pattern like Eustenopteron. The vertebrae show a mixture of fish and amphibian traits. In the caudal region, the fin rays of the fish tail are found. The girdles, limbs, and feet, however, are amphibian in structure and capable of locomotion on land. One finds this structural mosaicism as a characteristic of the evolutionary process.

The transition from water to land that occurred for vertebrates at the end of the Devonian created a new set of environmental challenges and influenced the importance of new structures. There was the challenge of respiration, where oxygen now needed to be obtained from the air rather than from the water. A transition from gills to lungs is a remarkable evolutionary event. Lungs, however, did not suddenly appear when vertebrates took to the land. The fish ancestors to the amphibians already had lungs, as well as gills. Frogs go through a dramatic development in which during the tadpole stage they live in water and breathe by gills, while during the frog stage they breathe air by means of lungs. Many species of fish have a swim bladder used as a hydrostatic organ and as a modified lung. There was the challenge of locomotion. Water is 800 times as dense as air. In water, the fish could be thrust forward by means of powerful muscles attached to a tail. Adaptation to water required a streamlining such as is found not only in fish but also in those marine mammals of land ancestry, the whales. The less dense air required a body-support system which involved less tail and the development of limbs.

In order to grasp the concept of water-land transition, one needs to observe not only fossils but also species that are alive today. Mudskippers are fish that are a few centimeters in length and live in mangrove swamps. Here, they feed upon insects and other invertebrates. While on land they are able to absorb oxygen through a moist skin and extract oxygen from water that is held in the mouth. Mudskippers can move on land by means of their pectoral fins which are fleshy and have an internal bony support. Another fish, the bichir, lives in rivers and swamps of Africa. It has both gills and a lung. It gulps air that is taken into the lung where oxygen is absorbed. Another fish, called the lungfish, lives in swamps. Periodically, when these swamps dry up, the lungfish moves to the bottom and forms a burrow. They surround themselves with mucous. The lungfish have two lungs and are able to gulp air by means of tubes connecting to the surface of the mud. They can survive for months under these conditions. When the rains come and the swamps are reestablished, the lungfish move out of their burrows, into water, and once again use gills for respiration, although they continue to gulp air for utilization by lungs.

Frogs, toads, and salamanders are truly transitional water-land animals. It is because of this characteristic of living in both land and water that they were named "amphibia." Amphibians are not completely adapted to land. Reproduction and development take place in water. The lungs of amphibia while functional are not very efficient, in that they have a small surface for the absorption of oxygen. In land forms such as reptiles, birds, and mammals, there is a greater dependence upon lungs, and one finds that these organs have a greater absorbing surface per unit area. Amphibia also may absorb oxygen through their moist skin and through their pharynx. Frogs spend their early life in water as tadpoles. At this time they breathe by means of gills. Young tadpoles have no legs, but as development proceeds and they prepare for life on land, legs develop, lungs develop, and the gills are lost. Some species of salamander live in water when young and on land when mature. If there is a wet season with little dry land, these salamanders remain in the water, retain their gills, become reproductively mature and breed. Another salamander, the mud puppy, spends its entire life in water and has both gills and lungs. The axolotl lives in water and has gills throughout its life. If, however, it is fed with a thyroid extract, it will lose its external gills and develop lungs. Thus, the biologist has been able to change an animal that has gills into one that has lungs by means of a chemical. Amphibians are tied to the water and must live near the water. Alfred Romer, a comparative anatomist, looked upon the amphibian as a peculiar type of fish.

Amphibians to Reptiles

The earliest amphibian was found in Greenland in rocks of the late Devonian. Given a new environment and in the absence of competition, they went through a period of diversification and adaptive radiation.

They were the dominant land group until superceded by the reptiles. The highest spot in the evolution of the amphibians is probably the appearance of Eryops, a labyrinthodont, which was found in Permian sediments of Texas. Eryops was about 5 feet in length. From its dentition it must have been a carnivore. It probably competed successfully with some of the early reptiles during the Permian period.

In the lower Permian sediments found in Seymour, Texas, some small vertebrates were found that were intermediate in structure between the amphibians and the reptiles. It is from the group called Seymouriamorphs that this form, Seymouria, arose. Seymouria had a skull and teeth like the amphibians but a post-cranial skeleton like the reptiles. This mixture of some primitive structures like amphibians and some advanced structures like reptiles makes it impossible to classify them. At present it is not clear whether Seymouria should be placed with the amphibians or the reptiles. It would be of importance to know the kind of egg that was produced, amphibian or reptilian-like. Such eggs, however, are not available in the fossil record. For convenience in classification, Seymouria is placed with the amphibians. It is considered to be a member of a group that is evolving toward reptiles and is intermediate between labyrinthodonts and reptiles.

With the reptiles there are three important innovations. (1) There is the appearance of a land egg. The embryo developed on land and was nourished by yolk in a large egg and prevented from dessication by surrounding fluids and membranes. The appearance of the amniote egg during Carboniferous is one of the significant innovations of vertebrates. It was comparable in importance to innovation of jaws and of limbs. (2) There was an external skin that was dry and prevented water loss. (3) There was internal fertilization. These properties freed the reptiles from the water. They now could exploit other environments on land. And exploit they did! During the Mesozoic era the reptiles were the dominant animals. They showed a remarkable evolution into many reptilian species. Some reptiles were as small as chickens; others were as large as a house. Some were herbivores such as the large brontosaurus; others were carnivores such as the huge tyrannosaurus. Certain groups of reptiles returned to the sea and developed flippers and long necks, the plesiosaurs. Others developed wings and could fly, the pterodactyls. There were mammal-like reptiles and bird-like reptiles and fish-like reptiles. The reptiles were the dominant group of vertebrates for about 100 million years. About 65 million years ago, these dominant reptiles died out. The reason for this extinction is not as yet known. An active field of research today deals with the question of reptilian extinction. One of the current hypotheses is that a large meteor struck the earth and caused the formation and release of a toxic substance (irridium). Support for this hypothesis rests on the finding of irridium in certain layers of the earth's crust and on certain astronomical observations. Although the scientists involved in this research cannot agree at present on the significance of the observations, they do agree that they have the ability to inquire about the condition of the earth 65 million years ago.

The earliest reptiles are called cotylosaurs and their fossils were found in Carboniferous strata from Nova Scotia. They are important, because they are the "stem reptiles" from which not only all of the other reptiles arose but also the birds and the mammals. As noted above, the ancient reptilian line was extremely diverse and the present-day survivors of the line show this diversity in appearance. The turtles, crocodiles, snakes, and lizards are all of the same "kind," in that they are reptiles, but they certainly are different kinds of reptiles. To the biologist, adaptive radiation and diversification among the reptiles is a good example of evolution taking place. Whether you wish to consider this to be an example of variation within a kind or the emergence of new kinds is irrelevant.

Reptiles to Birds

In 1860, soon after the publication of Darwin's *Origin of Species,* some fossils were found in a limestone quarry in Solnhofen, Bavaria, that showed a mixture of reptilian and avian characteristics. These fossils were called Archeopterix (ancient bird). They lived during the Jurassic period and were a dramatic discovery in support of the idea of evolution in the Victorian society of England. Archeopterix is considered to be a convincing missing link between reptiles and birds by biologists. Some people state, however, that Archeopterix is entirely bird because it had feathers and could fly. If Archeopterix is intermediate between reptile and bird, one should expect to find this reflected in its structure.

1. Archeopterix had well-developed feathers comparable to modern birds. Reptiles do not have feathers. As was discussed in Chapter 8, both feathers and scales are formed from a substance called "keratin." It has been determined that feathers are modified scales.

2. Archeopterix had heavy bony jaws lined with teeth, like reptiles. Birds do not have jaws and teeth. Birds have a lightweight, toothless beak made of keratin.
3. Archeopterix had a long-feathered tail that is supported by a bony extension of the spine. In birds there is no bony support of tail feathers.
4. Archeopterix had solid bones like reptiles. Birds have hollow bones that are very thin.
5. Archeopterix had no keel to which to attach wing muscles. Reptiles also have no keel. Flying birds have a well-developed keel for flight muscle attachment.
6. Archeopterix had claws on its wings (forelimbs). Reptiles also have claws. Birds have no claws on their wings. However, the hoatzin bird of Venezuela has two claws on each wing early in life but later loses these claws.

A much larger list of comparison has been made (Ostrom, 1974). Enough information, however, has been given to enable you to determine whether or not Archeopterix is a 100 percent bird or a reptile or an intermediate type. Biologists consider birds to be "glorified reptiles." They think that birds have descended from a small carnivorous dinosaur. Were it not for their feathers, Archeopterix would have been classified as a reptile. It shows more reptilian than avian characteristics. Little is known about the evolution of birds because there are very few available fossils.

There are about 8,000 species of birds today. These range in size from large ostriches to small hummingbirds, in diet from carnivores to herbivores, from fliers to non-fliers. In the 150 million years that birds have existed, their evolution has been very slow.

Among the modern birds are the ratites of Africa, South America, and Australia. These include the ostriches, rheas, cassowaries, emus, and kiwis. These birds do not fly and they lack a keel for the attachment of flight muscles. Both flying and non-flying birds lack a keel at the time that they hatch from the egg. In flying birds, the keel develops after hatching, while in non-flying birds the keel never develops. The large birds, ostriches and cassowaries, which are as tall as a man, look like dinosaurs in shape if one forgets the feathers. They are quite capable by their size and strength to be formidable predators. At one time large and carnivorous birds existed. Fossils of diatryma, a bird that stood seven feet high, were found in North America. A fossil bird, hespornis, possessed teeth.

Reptiles to Mammals

The changes from reptile to mammal took place between the Carboniferous and Triasssic epochs. Paleontologists can follow gradual change in structure from fossils that were clearly reptilian in the earlier geologic time to clearly mammalian in later periods. Within the fossil history of a group of reptiles, called synapsids, one finds remains of animals in which it is difficult to decide whether they should be classified as reptiles or mammals. They are called mammal-like reptiles. The most advanced of the synapsids were the theriodonts which are closely related in structure to the first undoubted mammals. The precise line of theriodonts that led to mammals is hard to determine. This is because in different groups of theriodonts there is a different mixture of reptilian and mammalian structures.

Comparison of the skeleton and teeth of reptiles and mammals show differences that enable paleontologists to recognize differences in fossils. Colbert (1980) has listed a number of skeletal differences between reptiles and mammals, some of which are summarized below.

1. Occipital condyles are structures found on the skull by means of which the skull attaches to the vertebrate column. Mammals possess two, while reptiles possess one occipital condyle. Some advanced theriodonts had two occipital condyles in the skull.

2. Reptiles have a small brain case, while mammals have an expanded brain case. In the theriodonts one finds a trend to the larger brain case.

3. The cheek teeth of present-day reptiles are simple, peg-like structures that are used primarily for gripping. The cheek teeth of the mammals are more complex and are used for grinding, tearing, or cutting. Differentiated dentition is found for the theriodonts.

4. The jaw of reptiles is made up of several bones including a small dentary bone. The jaw of mammals is made up only of a large dentary bone. Among the theriodonts the dentary bone is enlarged. Certain bones (the quadrate and articular bones) that join the skull to the jaw in reptiles have become transformed into the incus and malleus of the middle ear. Reptiles possess only a stapes as an ear ossicle, while mammals have the malleus, incus, and stapes as sound-transmitting structures. There has been a change in function from reptiles to mammals for these bones. In theriodonts the quadrate and articular bones are reduced in size. There are present in these mammal-like reptiles two kinds of jaw articulation. One is reptilian-like consisting of the articular and quad-

rate articulation; the other is mammalian-like consisting of the dentary and squamosal articulation.

The jaws and teeth have been particularly useful in the study of the reptile to mammal transition. This is because many well-preserved fossils have been found. They include fossils of late cretaceous mammals from Wyoming; very advanced mammal-like reptiles from England; fossils from China and from South Africa; and fossils from Arizona. Teams of paleontologists work in these different parts of the world and have devoted much effort to analysis of jaws and teeth. Changes in the jaws and teeth of fossils from primitive reptiles to early mammals have been studied by investigators over the past 100 years. It is evident to paleontologists that among the mammal-like reptiles there has been a progressive increase in the size of the dentary bone and a decrease in the size of the post-dentary bones.

In addition to these fossil transition species, we can find live organisms that carry characteristics of both mammals and reptiles. Such an animal was found about 1800 by explorers in Australia. It was called ornithorhyncus (bird bill) because it had a structure shaped like a duck-bill. This "bill" is not a hard bill made of keratin but rather soft and therefore cannot be compared to a bird bill. Ornithorhyncus is about the same size as a rabbit. It has thick, fine fur, like mammals. It builds a nest and lays soft-shelled eggs, like a reptile. The young that hatch from these eggs obtain nourishment from glands located on the ventral surface of the mother. These glands are not localized mammary glands but rather they are diffused. They nevertheless produce milk. In ornithorhyncus, a cloaca is found; that is, the rectum and urogenital systems are voided through a common vent like in reptiles. The spiny anteater, tachyglossus, is another primitive mammal that has fur and also lays eggs. Both ornithorhyncus and tachyglossus show both reptilian and mammalian traits. Their skeletons are more like reptiles (and therapsid) than like mammals.

CONTINENTAL DRIFT AND PLATE TECTONICS

Within this century a startling idea has emerged regarding the structure of the earth and the fixity of the continents. It is as revolutionary as the idea of a round earth and of the rising of mountains from the sea. This idea, called continental drift, is that the continents were not always in their present positions relative to one another. These continents now separated by huge oceans were once part of a single land mass. When first proposed, this idea was considered to be ridiculous, but within the past 10 years it has gained wide acceptance among paleontologists, geologists, and oceanographers. If one wishes to follow the development and testing of a new scientific idea that is in the process of emerging, the concept of continental drift is a good one to follow in its historic development.

It is presumed that at one time, about 250 million years ago, a single land mass, Pangea, existed. Pangea consisted of a large southern supercontinent, Gondwanaland, made up of South America, Africa, Antarctica, Australia, New Zealand, and India and a northern supercontinent, Laurasia, made up of North America, Greenland, and Eurasia. The present continents have drifted apart. South America and Africa, for example, are now separated by a newly formed South Atlantic Ocean.

The idea of continental drift goes back to at least 1756 to Theodor Lilienthal, who recognized that "the facing coasts of many countries, though separated by the sea, have a congruent shape so that they would almost fill one another if they stood side by side." Oceanographers have recently found that at the 3,000- to 6,000-foot level below the sea, the large land masses match up like a jigsaw puzzle.

During the period 1950-1970 detailed studies were made of the ocean floor. It was found that the earth's crust consisted of eight major and several minor plates. A large mid-oceanic ridge existed where the plates came together. At the borders of these ridges there were deep trenches from which there was an upwelling of volcanic magma to form more crust and to cause a lateral spread of the sea floor away from the ridges. The earth's crust is about 100 km in thickness. It rests on a fluid interior. The formation of more crust in one part of the world is compensated for by the loss of crust in another part by returning to the interior of the earth. Plate tectonics refers to the idea that the crust of the earth consists of plates. Because of this upwelling, the South Atlantic Ocean is becoming wider and South America and Africa are moving apart at a rate of 5 cm per year; North America and Europe have become separated to form the North Atlantic Ocean. Paleontologists are concerned with the distribution of organisms. They try to answer the question, "How did present-day organisms get where they now are?" They were impressed by the fact that similar rock formations and similar fossils often were found in widely separated parts of the world. The period covered by continental drift involves the later Paleozoic and Mesozoic eras. It was during this period that the tetrapod vertebrates were evolving. It is understandable, therefore, that continental drift is of particular importance to vertebrate paleontologists.

By comparison of their contours, Antarctica fits into the south-eastern part of Africa. If these two large land masses were at one time contiguous, one might expect to find similar organisms in rock layers of the same age in present-day Africa and Antarctica. Indeed, in 1968 a piece of a fossil jaw from a labyrinthodont amphibian was found in Antarctica embedded in coarse sandstone of Triassic, 200 million years old. How did this animal get to Antarctica? Could it have survived present environmental conditions? If one fossil bone was found, there must be others and therefore an expedition was organized and supported by federal funding to look for more fossils and to study Antarctica geology. In 1969, the paleontologists collected over 100 fossil bones of amphibians and therapsids (mammal-like reptiles). These fossils were like those found in South Africa. There was found also a plant fossil, Glossopteris. This plant had worldwide distribution and is characteristic of the Permian rocks of 200 million years ago. In 1970, Edwin Colbert made a significant find. He identified a fossil jaw of lystrosaurus and in subsequent digs found other such fossil bones.

Lystrosaurus is a "guide fossil." It is very abundant, it is easily recognized, it is confined to certain rock layers, it has a relatively short geological time scale. Lystrosaurus is a therapsid. It is about 4 feet in length, with a stocky build, short tail, no teeth, and beak-like jaws except for two large tusks. Lystrosaurus existed about 220 million years ago in Triassic rocks. In the Great Karroo of South Africa, abundant lystrosaurus fossils were found. Lystrosaurus also was found in India and in China. The discovery of lystrosaurus in Triassic rocks in the antarctic mountains was taken as proof of the concept of continental drift. It did not seem plausible to the scientists that these organisms could have migrated to Antarctica over thousands of miles of saltwater, nor to have been able to survive the harsh environment of the antarctic. One can also speculate that the temperature of the antarctic 200 million years ago was much warmer than it is today. Subsequent search for fossils uncovered certain antarctic vertebrates that were found to be associated in Africa. A prominent one is thrinaxodon, a small carnivorous, mammal-like reptile.

Correlation of Parts

George Cuvier (1789-1832) was a French biologist who established the field of comparative anatomy. He was interested in a principle called the "correlation of parts." The structure and organs found in an animal bear certain relations to one another. When we find a feather we know that it was part of a bird; we know that the bird had wings, two legs, and a type of musculature needed for flight. Cuvier could reconstruct an entire bird from a piece of leg bone.

The correlation-of-parts principle was applied to the study of fossils, and observations of many fossil forms led to the science of paleontology. Although Cuvier established the fields of comparative anatomy and paleontology, two disciplines that we consider crucial to the study of evolution, he was a firm believer in the fixity of species. He thought that there had been many great catastrophes that caused the extinction of many forms of animals. He thought that the last catastrophe was the Flood recorded in Genesis! He denied the existence of fossil man (Singer, 1941).

* * * *

A Charge

By all means, visit some of the museums of natural history. Study the exhibits. Keep going back! The museum literature is written in simple and understandable language. Museums are pleasant places to learn. Compare the bone structure of a wide array of vertebrates.

CHAPTER 11

THE EVOLUTION OF MAN

Cultural Evolution

PEOPLE did not suddenly appear on earth about 6,000 years ago as herders of sheep, keepers of gardens and with a language. During their early life people lived like animals. It was only after many years of biological evolution that they acquired the physical and mental equipment that was necessary to undergo cultural evolution, the transmission of ideas and tradition. The biologist Dobzhansky has stated, "Creation is not an act, but a process." The study of the origin of people is a particularly emotional subject, because it contradicts the creation stories put forth by many different religions. The reconciliation of religious and scientific views of creation is a major psychological problem for those who accept the first chapters of Genesis as historic rather than spiritual truth.

Historians describe the past by the analysis of written records. Since writing was invented about 3,000 B.C., there is a limit to how far into the past the historian can go. There are, however, other kinds of records, the physical remains of ancient villages, that enable another kind of historian, the archeologist, to describe the past history of people. About 1850, a Frenchman, Jacques Boucher de Perthes, looked for prehistoric flints in the Somme River Valley. He found stone tools in the same layers of earth as extinct species of animals and concluded from this that mankind was very old. The cultural history of people is usually divided into a long Paleolithic stage of over three million years, during which they were hunters and food gatherers, and a short Neolithic stage during which grain was grown from seed and livestock was domesticated. The discovery of agriculture, about 7,000 B.C., was a major achievement in cultural development. It enabled people to live in larger groups, to exchange ideas, and to develop systems of government and communication.

Work by anthropologists over the past 100 years has shown that people existed and had a culture much more than 6,000 years ago. Various prehistoric sites have been discovered where people have lived together. Caves existing 14,000 years ago in Altamira, a region on the coast of Northern Spain, have shown paintings and sculpture and evidences of a religion. Another prehistoric site in Czechoslovakia was inhabited about 25,000 years ago and revealed the presence of tent-like huts, animal skins, tools, musical instruments, artwork, and graves. Students of anthropology agree that a hominid creature that was physically indistinguishable from us existed about 35,000 years ago. He is considered to be a member of our species, *Homo sapiens*. Studies of various sites inhabited by these "Cro-magnon" people showed that they possessed the following cultural characteristics. They could make and use fire; they made jewelry; they decorated their bodies with tatoos; they made musical instruments; they painted on the walls of caves; they had graves for their dead.

The evolution of culture is thought to have begun over three million years ago with a hominid that had bipedal locomotion and hands that could use stones as tools. This hypothesis is based on the finding of both hominid bones and tools in the same sites. Over this period of three million years there has been an improvement and increase in the sophistication of the tools that were made from stones and from the bones and horns of other animals.

Evidence for the use of tools have been found in sites (beds) in Lake Turkana, Olduvai Gorge, and Hadar in East Africa. These beds contained stone tools that have been dated to about 2.5 million years ago. The most complete demonstration of stone tools has been made by Mary Leakey from the beds in Olduvai Gorge by the Habilines. These tools include choppers for the butchering of large animals, stone balls that were used as hammers, scrapers for the cleaning of meat and fat from animal hides, chisel-like tools for shaping wood, sharp flakes of stone for cutting. Various beds have been studied and been shown to have a pattern of human-like activity. Some beds, 10-feet-by-12-feet square, are surrounded by rocks that are piled into heaps to indicate a shelter. Other beds seemed to serve as butchering sites, since scattered bones of elephants and butchering tools were found. It seems clear from these studies that these early Pleistocene hominids made simple tools according to recognizable pattern; that this way of making tools became a tradition that was passed on to other generations.

A summary of the evolution of culture in man shows a list of impressive accomplishments. These accomplishments start with the use of

TABLE IV

EVOLUTION OF CULTURE IN MAN

Type of Hominid	*Years B.P.*	*Cultural Era*	*Activity*
Australopithecines	3-1.5 million years	Paleolithic Lower	used tools for hunting and butchering
The Habilines	2-1.5 million years		probably made tools
Homo erectus	400,000 years		used fire
Home neanderthalensis	100,000 to 40,000 years	Middle	used ceremonies and burials for the dead
Homo sapiens (Cro-magnon)	35,000 years	Upper	made music, painting and sculpture; probably developed language
	10,000 years	Mesolithic	
	7,000 years	Neolithic	start of agriculture food raising and animal husbandry
	5,000 years		start of Judeo-Christian religions
	500 years		development of modern science
	200 years		the industrial revolution
	15 years		man on the moon—the space age

tools, to the use of tools to make tools, to fire, to ceremonies, to art, to language, to agriculture, to religion, to modern science, to the industrial revolution, to the space age. These accomplishments have required a great deal of intelligence and involve a learned behavior. One must be intelligent to solve problems by the use of tools and to adapt to new environments such as flying in the air and swimming underwater by means of machines. One must be intelligent to recognize spiritual concepts and to have a sense of the past and the future. We can look with pride at our cultural development. It was not given to us. We earned it. There is a lesson here to uplift the human spirit rather than to depress it.

William Shakespeare stated that, "Some men are born great, some achieve greatness, and others have greatness thrust upon them." If you had a choice, which alternative would you choose?

Human Evolution—Approach and Attitude

In 1925, Raymond Dart, a professor anatomy, reported the discovery of a skull that he had found in South Africa. It was clearly a hominoid, but there was a question as to whether it was ape-like (pongid) or man-like (hominid) in appearance. This skull was about two million years old. Dart named it *Australopithecus africanus*. Since that time, other fossil discoveries have been made by other paleoanthropologists who now suggest that during the Pleistocene era there were three or four species of *Australopithecus* that lived in Eastern Africa. At the present time, intensive investigations are being made at over ten different sites from South Africa to Ethiopia. Thousands of bones, bone fragments, and teeth have been found. These sites are laboratories for scientists from all over the world to study fossils and to test ideas on one another.

The hypothesis that *Homo sapiens* evolved from non-human ancestors is a testable hypothesis and therefore falls under scientific investigation. There is a limitation to our studies, in that we may never be certain how two groups of organisms are related because we cannot go back in time to see the evolution of one group into another. We do not, however, throw up our hands and say that the problem is insoluble. We have here a "black box" where we try to describe events that no one can see. This, however, is the usual situation that scientists are in. We can study the structure of fossils; we can determine the age of fossils; we can determine whether the fossils are more similar to apes than to man; we can study the changes in structure that occur with time and determine whether the older fossils look more or less like modern man than do the younger fossils. We can study the changes in culture that have taken place among hominids with time and determine the uniquely human activities in fossil people. We can inquire concerning at what points in time ancient hominids can be classified as humans.

After Dart's discovery in 1925, those people who were competent in the field of primate anatomy and who were intensively involved in this specific research area studied the bones and teeth of australopithecines and argued among themselves for about 25 years. During this period more bones and teeth of *Australopithecus* were found. After careful, systematic, and self-critical analysis, the issue was settled in the minds of this scientific community who concluded that *Australopithecus* was a hominid and was a forerunner of *Homo sapiens*. This conclusion was based upon the facts available at that time. If later data show this conclusion to be wrong, it will be discarded or modified. It has been said that people who never make mistakes never make discoveries. In science, mistakes in observation and interpretation occur and we sharpen our concepts from our mistakes. The attitude of religious fundamentalism is that it makes no mistakes in observation or interpretation.

How does the anthropologist determine whether *Australopithecus* is man-like (hominid) or ape-like (pongid)? LeGros Clark approached this problem by studying the structure of present-day apes and comparing them with present-day man. In this way he could determine which characteristics were uniquely pongid and which were uniquely hominid. He then examined the bones and teeth from the fossil australopithecines and compared them with the comparable structure from apes and men. He concluded that *Australopithecus* was a hominid.

Throughout the world there are hundreds of museums of natural history. These museums are used for research and for the education of school children and adults. If one wishes to see the evidence for evolution, one should visit these museums. Each museum of natural history is a monument to the idea of evolution.

The American Museum of Natural History in New York City had an exhibit (April 13 to September 19, 1984) called "ANCESTORS, Four Million Years of Humanity." This exhibit consists of the original fossils that document the long course of human evolution. The exhibit was put together by paleoanthropologists from different parts of the world.

If one studies this exhibit, one can see the orderly sequence of structural changes with time in human evolution. The concept of human evolution as an historic process is not under debate. All paleoanthropologists agree that man evolved from an ancestor that was not human. Although authorities can be wrong, I am more inclined to accept the opinion of people who have studied human evolution over people who have not, or dare not. What is also impressive in the exhibit is the care with which statements are made and the conscious attempt to assess the limits of accuracy of conclusions.

The Ancestry of Modern Man

Man appears to have been derived from the australopithecines which in turn may have been derived from ramapithecus, a creature that lived 10 million years ago. The australopithecine showed from its bones and teeth a mixture of ape and man characteristics. The skull and face were ape-like, the dentition was man-like, and the pelvis was man-like and showed upright locomotion. The changes from *Australopithecus* to modern man that have occurred over a period of two million years are as follows.

The brain case became larger to house larger brains. This is correlated with increased intelligence. A vertical forehead developed. The eyebrow ridges became reduced in size. The jaw became shorter and the teeth became smaller. The face became flatter. These changes have been progressive, in that the fossil men between the australopithecines and modern man have shown intermediate physical characteristics, and more recent fossils are more like modern man than more ancient fossils. Correlated with these physical changes are the progressive cultural changes that have occurred. More detailed analysis will be found in all physical anthropology texts.

As a teacher, it is my responsibility to present this current scientific view and its limitations concerning the origins of people to my biology classes and to do this regardless of my personal convictions. If I disagree with the conclusions reached by the specialists in primate anatomy, I must first learn primate anatomy and how to analyze the bones and teeth. I need to make myself competent enough to express an opinion. I will need to become involved in discussions with this scientific community by publication in scientific journals and by speeches at scientific meetings. I will not be able to convince the scientific community of my views by merely presenting these views to church groups and to public school children, or by censoring textbooks. The issue as to whether and how people evolved is a scientific matter. The issue as to whether human evolution is dangerous knowledge for young people is another question.

The Chromosomes of Man and Chimpanzee (Chromosome Homology)

The diploid chromosome number for man is 46. These chromosomes show differences in length, position of a structure called the centromere, and the array of genes. Detailed comparison of the chromosomes became possible with the discovery in 1970 of techniques to recognize transverse bands in the shorter metaphase chromosomes where about 500 bands could be seen. Analysis of the longer prophase chromosomes increased the resolution so that about 1,200 bands could be recognized per haploid set. One can see banding differences not only between chromosomes but within chromosomes. The trained microscopist can recognize not only the characteristics of each chromosome but of each part of each chromosome. The diploid number of chromosomes for the chimpanzees is 48. There are two shorter chimpanzee chromosomes that match up with one longer chromosome (chromosome no. 2 of man). This can be explained by a fusion of the two chimpanzee chromosomes to form the single chromosome found in people.

Since 1975, the mammalian cytologists have agreed that human and chimpanzee chromosomes have a considerable degree of homology. Analysis of the banding patterns was made by Jorge Yunis (1980). By means of a high-resolution chromosome technique it was possible to study band by band the chromosomes of man and chimpanzee. Yunis showed that "virtually total homology exists for bands in the two species."

In addition, it has been reported that more than 20 genes have been localized to homologous chromosomes and chromosome bands of the two species. Further, the clinical, behavioral, and cytogenetic features of Down's syndrome was found in a chimpanzee.

Further studies by Yunis (1982) compared the chromosomes of man, chimpanzee, gorilla, and orangutan. The three ape species have a diploid number of 48. The analysis showed that 18 of the 23 pairs of chromosomes are virtually identical. This is taken to indicate that there was a common hominoid ancestor.

The Nucleic Acids and Proteins of Man and Chimpanzee (Chemical Homology)

How similar are the nucleic acids and proteins of man and chimpanzee? King and Wilson (1975) addressed this question and found that the differences were small. This indicates that the genes of man and chimpanzee are as similar as those of closely related species of other organisms.

A comparison of proteins with regards to their amino acid sequences showed that their alpha hemoglobin, a protein consisting of 141 amino acids, was identical. For 12 homologous proteins with a total of 2,633 amino acids there were only 19 amino differences ($^{19}/_{2633} = {}^{7.2}/_{1000} = .007 = < 1\%$). The probability that alpha hemoglobin of man and chimpanzee would be identical in amino acid sequence if these species were unrelated would be $^{1}/_{20^{141}}$. This is such an incredibly small chance that the molecular biologist considers it more probable that their identical structure is due to common ancestry.

Although man and chimpanzee are 99 percent alike in their proteins and DNA, they clearly are different in appearance. If they have the same genes, why don't they look alike? There are two classes of genes. There are structural genes that determine the kinds of enzymes, hemoglobin, and other proteins that are made; such genes determine the amino acid sequence of proteins. There are regulatory genes that determine when and for what duration the structural genes are turned on. Genes are turned on and off during development. For example, the gene for the synthesis of the gamma chain of fetal hemoglobin is made only before birth, while the gene for the synthesis of the beta chain of adult hemoglobin is made after birth. The period of this structural gene activity is monitored by regulatory genes. The chimp and man may be different because of a few regulatory gene differences, and these control the timing of protein synthesis and the quantity of protein synthesis. Further, a single regulatory gene may activate a group of structural genes. The small modification of a single developmental event can cause large changes in the subsequent development of the embryo, because many later events can be dependent upon the earlier event.

CHAPTER 12

THE ORIGIN OF SPECIES

SPECIATION

WHAT IS a species? Are species real (natural units), or are they imaginary units living in the minds of biologists? These questions have bothered biologists for over 200 years. Charles Darwin wrote about *The Origin of Species*. The initial shock to Darwin and to nineteenth century biologists was the discovery that each species arose from other pre-existing species and was not created separately out of nothing.

For populations that reproduce sexually, a species is a population in which its members can share genes through reproduction. The continued existence of the species is dependent upon the reproductive interdependence of its members. Dobzhansky has stated, "Thus Mankind is more than a name for a class of objects. It is a biological reality; all human beings are part of the gene pool from which the genetic endowments of their posterity will arise. A sexual species is the ultimate extention of the ancestral family, it is also the family of the future." All people belong to the same species, *Homo sapiens*.

Thus, populations that can reproduce and produce viable, fertile progeny belong to the same species, while populations that are reproductively isolated belong to different species. To the biologist who is concerned with the classification of organisms, it is frequently difficult to use reproductive isolation as a criterion of species. There are so many different species that one cannot attempt to make all of the crosses. There are therefore many geographically isolated populations that look very much alike that have been classified as different species because they have no opportunity to interbreed. Whether they could interbreed has not been tested. It is often not worth the time and effort to determine by means of mating experiments whether two populations belong to the same species or to different species. One cannot determine whether species that existed in the past are the same species that exist today. One cannot cross a modern alligator with an ancestral alligator. The term "chronospecies" is used to refer to ancient organisms that look like modern ones but cannot be tested to determine their affiliation to modern species.

Species are usually classified by their appearance (morphology). Originally, it was assumed that every member of species looked like every other member. Thus, if one were to describe one beetle of a species (the type specimen), one would have described all the beetles of that species. Whenever a variant beetle was found, it was considered to be an imperfection and was discarded. Whenever a new kind of beetle was discovered, it was compared to the type specimen beetles in the museum and was placed into the same species or a different species depending upon the judgment of the classifier (taxonomist) concerning the closeness of their morphology.

One difficulty in identifying species in this way is that there are some populations that look so much alike that they cannot be distinguished by their morphology. These are called sibling species and, despite their similarity, they do not interbreed and exchange genes. They are thus different species.

Modern taxonomy has discarded the type specimen idea. A species is a population of organisms in which there is a great amount of variation. No single organism is characteristic of its species. The amount of variation within a species may be greater than the mean variation between species.

The lesson of evolution is that a species is not static and unchangable. It is a dynamic unit that responds to its different environments and changes over the ages. If there were a constancy (fixity) of species as the Swedish taxonomist Linnaeus had believed, the classification would be much easier. Since species change and grade into one another, delineation into distinct groups becomes more difficult. Evolutionary biologists are less interested in the species (the product) than in speciation (the process). The question is not how many species there are, but how the species arise. The Jesuit priest and paleontologist Teilhard du Chardin described evolution as "the process of becoming and not the fact of being." Evolution to biologists is a process by means of which new and diverse species arise. Because speciation is often a gradual process, a species is often difficult to delineate. It has been stated that if one cannot define a species, one does not know what it is or whether species are real biological units. One can clearly recognize, however, that cows, cats, and dogs belong to different species. Try to define love, pain, God, and decide whether they are real. Two kinds of speciation have been recognized by biologists. One is called *geographic speciation*. This is a gradual change in the genetic constitution of populations so that over a period of thousands of generations they become reproductively isolated. The other is called *quantum speciation* because it can occur within one or a few generations.

Geographic Speciation

This type of speciation usually starts because of geographic isolation. A species will spread to new territories and thus the more distant populations are reproductively isolated from one another. Or, barriers between populations may arise such as mountain ranges or water. Over a period of many generations, these reproductively separated populations adapt by natural selection to their own environment. This results in genetic changes and thus to differences from one another. These changes that occur are so gradual that one is unable to perceive them from one generation to another. At first the separated populations are considered as different races (subspecies). If the barrier is removed, races that occupy different territories (allopatric races) may come together into the same territory (sympatric races). Interbreeding may then occur and these races may fuse into a single species. Thus, the genetic divergence of populations initially leads to differences that can be reversed. Some isolated populations, however, accumulate so many genetic differences that they no longer are capable of interbreeding to produce viable or fertile progeny.

It appears, therefore, that a gradual gradation of races into species occurs in space and time. Races are often incipient (beginning) species. Over a period of reproductive isolation, genetic differences between populations become so great as to prevent interbreeding. These changes are so small that one cannot point definitely to a precise moment when two populations are not two races but two different species. Is there any evidence in support of this intergradation between races and species?

The herring gulls are represented by two reproductively isolated groups that live in Northwest Europe. They are sympatric. They do not hybridize and thus behave as separate species, named *Larus argentatus* and *Larus fuscus*. These species, however, are connected by a circle of intergrading races living in North America and Northern Asia. They live around the Polar Ocean in Labrador, Canada, and Siberia, Northern Russia. Gene exchange occurs in these populations but not in Northwestern European populations. Thus, the populations act like races in one part of the world and species in another.

The leopard frog, *Rana pipiens,* has a distribution range from Northern Canada to Central America. Morphological differences exist for populations from different geographic regions. These populations can be considered as subpopulations of the species. The distribution from north to south, however, is continuous and thus the delineation into races is artificial and not needed. Crosses between adjacent populations anywhere in this range give rise to viable, fertile progeny. They thus appear to act like the same species. Crosses between populations at the extremes of range (frogs from Canada crossed to frogs from Central America) give rise to inviable embryos. It is clear that these frog populations have adapted to temperature differences in the northern and southern regions. Frogs from nothern populations will develop at lower temperatures, but they cannot develop at the higher temperatures required by the southern populations. Crosses between the extremes in frog populations do not hybridize and therefore act as bona fide species.

An elegant series of experiments to demonstrate the speciation process was performed in a *Drosophila* group by Ayala (1975). This group consisted of 15 species, 6 of which were sibling species. These species have a wide distribution in the tropics of South America and range into the Caribbean region. One has never observed a change from one species to another among these *Drosophila.* If, however, the speciation process is taking place, one should be able to determine that certain groups are more closely and other groups are less closely related. The degree of relationship was shown by determining the extent of reproductive isolation and the extent of genetic difference by analysis of protein variants. Samples of *Drosophila* populations from different parts of South and Central America were collected. Their reproductive and mating patterns were studied. Different degrees of evolutionary divergence were demonstrated. This is the pattern to be expected if speciation is occurring in time and space.

Islands represent the ultimate form of geographic isolation. Islands are relatively small bodies of land cut off from major land bodies by bodies of water of varying dimensions. If islands are formed by the breaking away of a portion of a pre-existing mainland, the flora and fauna of the island will be closely related to the nearby mainland.

Islands, however, frequently are formed *de novo,* either through volcanic action or as coral reefs. In this case, the island will not have a native population of living beings, but rather one will develop over a protracted period of time via colonists brought to the island by wind or water action from other places. Once on the island, the newly introduced colonist can begin to live in whatever niche of the island it has attributes for. Thus, islands provide laboratories of a multitude of natural experiments to study the forces involved in the evolutionary process. Two of these laboratories are discussed in greater detail below.

Darwin's Finches, Variations on a Theme

The Galapagos Islands lie in the Pacific Ocean about 600 miles from the coast of Ecuador. To the west of this cluster of islands there is no land for 3,000 miles. These islands arose out of the sea due to the activities of volcanoes under the ocean. The hot magma from the interior of the earth cooled and solidified to become rock that in some areas had been built up to 5,000 feet in altitude. Upon these initially barren rock formations one finds a range of plants from cactuses in arid areas to large trees in more humid regions. Certain reptiles are found (the land and the marine iguana, a giant land tortoise). There are about 80 species of birds. The Galapagos Islands are of historic significance to biologists, because it was here that Charles Darwin made observations that led to the concept of evolution.

On the Galapagos, Darwin recognized a group of finches. These groups were morphologically distinct and yet were clearly similar. There were 14 species of finch that were related to one another in beak structure, short tails, body form, and plumage. They showed variations in their beaks, habits, and diet. Darwin, in seeing this diversity among the finches, remarked, "One might fancy that one species had been taken and modified for different ends." A cluster of islands that initially was barren and now supports different species of plants and animals raises some questions. How did these different species get to these islands? Why are there very similar but distinct species on these islands? Why do these different species occupy different habitats and eat different food? How do they come to possess structures and behaviors that adapt them to specific niches?

The Galapagos Islands were initially inhabited through migration by a few organisms of a few species. These species resembled those found in South America. These migrants established themselves, overpopulated their habitat and extended their range to different parts of an island from one island to another. They exploited new environments and became reproductively isolated from populations living in different environments. One recognizes among the finches that there have been dramatic modifications that relate each population to its environment. This type of diversification is called *adaptive radiation.* It occurs more rapidly in island populations in which there is no competition from other species and there are no predators. This adaptive radiation could not occur where there is a great amount of competition. No such wide range of finch variations therefore is found in South America.

Darwin saw different forms of finch on the Galapagos Islands. Were each of the 14 species created separately, or did they arise through modification from a few finch ancestors? Darwin concluded that descent with modification had occurred.

Darwin was fully aware of the impact of his theory on biblical literalism and on his own religious thinking. He wrote, "At last gleams of light have come and I am almost convinced (quite contrary to the opinion I started with) that species are not (it is like confessing murder) immutable." Once Darwin recognized that there were serious conflicts between his scientific views and his religious views, what should he have done? Should he have discussed his data and interpretation with the clergy to determine whether his ideas were valid and compatible with Christian thought? How and when are new scientific ideas to be presented?

Hawaii, Showcase of Evolution

Hawaii has eight major islands that collectively are equal in area to Massachusetts. These islands are of volcanic origin. They emerged out of the sea. There has been great interest in Hawaii because it shows such dramatic manifestations of evolution. Estimates of the age of these islands range from 40 million years for Emperor sea mounts to Hawaii which is nearly 700,000 years old. The oldest island has a summit of 5,000 feet, while the youngest island has a summit of 14,000 feet. These islands show considerable environmental diversity not only in altitude but also in humidity and temperature. There are areas with luxuriant vegetation, deep canyons, rivers, sand beaches, and forests. In addition, they are isolated from the continents by thousands of miles and therefore it is difficult for organisms to get there. There are no large mammals, no reptiles. The diverse species on these islands arose from a chance arrival of individuals (founders). Most of the plants and animal species native to the islands are found nowhere else in the world. The Hawaiian Islands are an exceptional natural laboratory for the study of evolution. Hampton Carson, an evolutionary biologist, and colleagues are studying how this process operates, how the genetic constitution of populations changes to adapt to new environments. Some of these studies are described in *Natural History* (1982).

Some forms such as the upland goose have remained as a single species, but for others considerable diversification has occurred. Some of the small moths have radiated into a thousand new species. *Drosophila* flies have radiated into 600 different species. In this small area of Hawaii about 40 percent of the known *Drosophila* species of the world are found. Further, their morphological, ecological, and behavioral diversity exceeds that of *Drosophila* in the rest of the world.

A group of plants, the silversword alliance, belonging to the sunflower family shows a remarkable radiation. A single species has diversified to occupy nearly every type of habitat in the Hawaiian Islands. The silverswords range in form from small mat-forming shrubs and ground-hugging rosette plants to climbing vines and even sizable trees. Some live in moist and others in dry environments. There are about 28 species of silversword that are close enough genetically to form hybrids. All species seem able to interbreed, but there is a high degree of hybrid sterility. For most of the species the chromosome number is 28, but for some species the number is 26. Since crosses can be made, the genes involved in the various adaptations can be identified.

The Hawaiian Honey Creepers (Drepaniidae) are a group of finches that have undergone remarkable diversification in bill structure and in feeding habits. These bills range from long and slender to short and stout, from straight to curved. Some have a long, curved upper bill and a short, stout lower bill. Associated with these bills are the feeding habits. Heavy bills used for cracking hard seeds or crushing dead twigs to expose beetle larvae. Very long, slender, and strongly curved bills can probe into crevices in the bark of trees and to probe flowers. Some honey creepers have a heavy lower mandible to break away bark and a long curved upper mandible to sweep through the crevice. All of these species have apparently evolved from a common finch-like ancestor.

Quantum Speciation

Geographic speciation is a slow process in which the formation of new species requires thousands of generations. Quantum speciation is sudden and new species can be produced in one or a few generations. This sudden speciation occurs because of chromosome changes involving either polyploidy (chromosome doubling) or chromosomal rearrangements. Polyploidy is particularly common among ferns and flowering plants. Almost one-half of the species of flowering plants have polyploid species. Among the wheats there are 2N, 4N, and 6N species; in cotton there are 2N and 4N species; in mints there are 2N and 4N species. There are bananas that are 3N; potatoes that are 4N; strawberries that are 8N.

In the mint family there are two diploid (2N) and one tetraploid (4N) species. *Galeopsis speciosa* and *Galeopsis pubescens* each have a diploid number of 16, while *Galeopsis tetrahit* has 32 chromosomes. A cross of speciosa × pubescens produced a sterile hybrid (8 speciosa and 8 pubescens chromosomes). A doubling of the chromosome number, however, produced plants with 16 speciosa and 16 pubescens chromosomes. These plants, now fertile polyploids, looked exactly like the *G. tetrahit* found in nature. When these laboratory-formed hybrids were crossed to the natural tetrahit, fertile and viable plants were produced.

The cultivated cottons *(Gossypium)* found in the United States have 52 chromosomes (26 pairs). There are two wild cotton species, each containing 26 chromosomes (13 pairs). In one, the American wild, the chromosomes are small and in the other, the Old World wild, the chromosomes are large. The cultivated cotton has 13 pair of chromosomes like the Old World and 13 pair like the American cotton.

Among the wheats *(Triticum)* there are different species that fall into 2N, 4N, and 6N groups. The diploid einkorn wheats have 14 chromosomes, the tetraploid emmer wheats have 28 chromosomes, the bread wheats have 42 chromosomes. Could it be that the bread wheat arose from a cross of diploid and a tetraploid species? The emmer wheat (tetraploid) was crossed to a diploid wild grass, Aegilops, and the chromosome number of the hybrid was doubled by means of a chemical, colchicine. This hexaploid produced functional gametes. This laboratory-made wheat was crossed to a naturally occurring bread wheat and produced fertile hybrids. This proved to investigators that the natural bread wheats arose from a cross of the emmer wheat and the Aegilops.

Quantum speciation may arise also by means of chromosomal rearrangements. Translocations may occur in which parts of chromosomes fuse with other chromosomes. In this way the chromosome number can change. Crosses between individuals that have different chromosomal rearrangements results in progeny that are highly sterile and in the establishment of different groups that breed among themselves. The pocket gophers of the *Thomomys talpoides* group are found in the Rocky Mountains of the United States. Different isolated populations have chromosome numbers that range from 40 to 60. These gophers from different regions are morphologically alike. Among groups of animals that have a limited mobility, new species can thus arise.

The Diversity of Species

Suppose that you wanted to create an earth and populate it with plants and animals for your use? What kinds of organisms would you choose to create? You would make fruit trees, vegetables, and cereals to furnish food; flowers and shrubs for pleasure; horses or camels; cows, goats, and sheep; chickens and ducks; dogs and cats; and fish, lobsters, and oysters. Unless you decided upon other sources of energy, you would need to provide food for your creations. Plants would require photosynthesis; herbivores would eat plants. Carnivores, however, would require fresh meat from slaughtered animals. A large energy problem would be solved by resorting to photosynthesis. This would eliminate the 90 percent energy loss with each organism in the food chain and the unpleasantness of animal slaughter.

Would you create rats and snakes? Would you create spiders and wasps, sharks, starfish, horseshoe crabs? Would you make human parasites such as plasmodium (for malaria), entamebae (for dysentary), trypanosomes (for sleeping sickness), hookworms, tapeworms, intestinal bacteria, viruses? Would you make poison ivy? For every species that you created you are obliged to state the purpose for its existence and its use to mankind.

It is clear that few kinds of organisms would be needed; perhaps 100 species could satisfy human needs. How many species, however, exist on earth? The total is not known, since new species are continuously being discovered. The estimates range from 3 million to 30 million. If one is going to justify the existence of a species by its usefulness to human beings, how is one to account for the millions of species that exist at the present time? How does one explain the existence of 10,000 species of sponge? And, how is one to account for the countless extinct species? In what way did dinosaurs serve man? What is the meaning of this enormous diversity?

When one examines a partial list of species, one finds a rather curious mix, in that there is a wide divergence in the numbers of species in different groups. Over 900,000 species of arthropod have been named but only about 40,000 species of vertebrate; 100,000 species of mollusc and 8,000 species of annelid worms. The phylum Arthropoda consists of five major classes: Crustacea, Chilopoda (centipedes), Diplopoda (millipedes), Arachnida (spiders), Insecta. The Insecta consist of about 25 orders, including Hymenoptera (bees, wasps, ants), Lepidoptera (moths and butterflies), Diptera (two-winged flies), Coleoptera (beetles). About 40 percent of all species in the phylum Arthropoda are beetles.

It is interesting to inquire where organisms live. Some are found exclusively in caves; some in the depths of the ocean. Some organisms live in or on other organisms; in potatoes, in frogs, in termites, in bacteria. Some organisms live in the arctic, some in desserts.

One can see the diversity of organisms and their bizarre forms and structure by visiting an aquarium or a zoo, or by taking a walk in the woods or field. Collect some pond water, filter most of the water through sand or filter paper, and examine samples under the microscope. You may see worms or insects, or algae or protoxoa. By such microscopic observation, thousands of species have been identified.

If only a few "kinds" had been created in the beginning, about 6,000 years ago, how does one account for the millions of species that exist today? If only one or even a few kinds of snake were originally created, how does one account for the formation of 2,000 species of snake in only 6,000 years? For what purpose have 2,000 species of snake come into existence?

The existence of millions of species is explainable by evolution but not by creationism. The concept of evolution involves changes in the structure of organisms through time, the transformation of one type into another, and diversification from common ancestry. It involves the formation of many present-day species from a relatively few ancestral species. The creationist view also must account for the formation of millions of species from a few ancestral "kinds." If the term "kind" is to be used in a scientific way, it must be more precisely defined. In a rather curious way the major differences between the biblical account and the evolutionary account may turn out to involve the duration and rate of speciation. In the evolutionary account there has been a slow and imperceptible speciation over millions of years. From the biblical account it can be inferred that a rapid speciation has occurred over the short period of 6,000 years. This rate of speciation would be so rapid that if it occurred, it should be clearly observable. The loss of limbs by the original snake in the Garden of Eden and the production of limbless forms from an "unlike ancestor" could be used as an example of evolution.

Notice that the reasoning and approach of Ayala in dealing with speciation in *Drosophila* is similar to the "picture frames" used for mitosis and to the sequence of changes with time for the horse evolution. How does one measure changes that are imperceptible during one's mere lifetime?

CHAPTER 13

THE ORIGIN OF LIFE

WHEN SPEAKING of the origin of life, the quip is sometimes made that "half of the hypotheses are correct and half are wrong, but we do not know which half is which." This quip does not embarrass the scientist. He often speculates. This is the way that scientific laws begin and knowledge is accumulated. Someone gets an idea and it is tested by experiment. An idea may be ahead of its time because tests are not available or because it does not fit into our conceptual framework. The idea then sits in limbo. Suppose that our current hypotheses are wrong? So, what then? They will be replaced by other hypotheses that give a closer approximation to the truth. This finer tuning of concepts is never completed.

The origin of life is being actively investigated today. The goal is to discover the design (the mechanisms) that explains how life was created. Eventually, this research will enable us to make living organisms and complex proteins from non-living substances. The scientific research does not inquire into the purpose for which life was created. It has no bearing on the existence of a Creator or whether God intervened. To the scientist, the origin of life is not a moral question any more than the rotation of the earth around the sun is a moral question. To the citizen, the search for the origin of life should be a quest, an exciting adventure. It is a measure of the intellectual potential of *Homo sapiens*. We should be proud and not fearful of this quest.

The twentieth century is an exciting time for the citizen with an open mind. Physicists are finding the elementary particles of which atoms are made. Geographers are studying the drifting apart of continents. Oceanographers are discovering new and bizarre organisms on the ocean floor. Astrophysicists are investigating the origin of the universe (about 20 billion years ago). Astronomers are using infrared light to discover millions of new stars. They estimate that 30 percent of these stars may have planets rotating around them. This is not an age to be bound by a presumed immorality of scientific ideas. There are sufficient moral problems in the societies of people and nations to challenge the efforts of all religions. The challenge of religion is to find how people are to live together as a human family. What is required is not a chauvinistic love but rather tolerance.

EARLY LIFE

The earth is about 4.5 billion years old, but it was not until 600 million years ago, at the start of the Cambrian era, that many different kinds of large multicellular marine organisms appeared. It is from these organisms that present-day plants and animals have descended. For over 100 years the study of evolution has dealt with those species that existed during the past 600 million years. It seemed that complex organisms had appeared suddenly in Cambrian. There was no evidence for Precambrian life.

It did not seem reasonable to the scientist, however, that these multicellular organisms could have appeared suddenly on earth by spontaneous generation and without ancestors, and yet he was unable to go back further in the fossil record into the Precambrian era. The skepticism for spontaneous generation (abiogenesis) was based upon early experiments that were done by Francisco Redi in 1688 on houseflies, Lazzaro Spallanzani in 1765, and by Louis Pasteur in 1862 on microorganisms. They showed that life can arise only from preexisting life. This biogenetic law is a basic principle in biology.

The skepticism for abiogenesis is based also upon our knowledge of the complexity of organisms. Complex organisms could not have arisen in an instant of time from non-living material. It is easily recognizable that a cat or a honeybee is a complex organism. The ameba, a single-celled protist, is also complex. The virus is complex. Each of the different components of a cell is complex. To the scientist this complexity did not suddenly and miraculously appear out of nothing but came about over a period of 3 billion years by a DESIGN whereby primitive microscopic organisms worked out by trial and error the biological and chemical strategies that made complex life possible. These simple organisms (prokaryotes) developed the patterns for utilizing energy, for feeding, for movement, for sex, and for the perception of the environment. Primitive cells worked out the basic mechanisms of fermentation, photosynthesis, respiration; the mechanisms of mitosis, meiosis, and fertilization. These processes are essentially identical in all organisms living today, from unicellular to multicellular, to plants, to animals. ATP for energy utilization and DNA for reproduction are used to explain these biochemical processes in present-day ameba and elephants and earthworms and people. The presence of these complex biochemical and biological processes in all organisms suggests that all life is ultimately related to one common ancestor.

Indeed, it was the development of the processes of energy utilization and of reproduction that made it possible for more complex organisms to exist and evolve. The evolution of these fundamental chemical reproductive features took more time to develop and be selected for (3 billion years) than the evolution of fish to mammals (400 million years). Once a reproductive mechanism for the selection of favorable genes appeared, evolution occurred more rapidly.

In the evolution of organisms certain structures have appeared that are characteristic of certain groups of organisms. All birds have feathers; all mammals produce milk for their young; all chordates have a notochord. These are called *derived structures* because they are not essential for life. Such derived structures vary considerably from one group of organisms to another. Recognize the marked difference between whales and frogs in these derived structures and the rapid evolution of these derived structures. At the biochemical level the processes of energy utilization and reproduction are similar in all organisms (including whales and frogs) and are therefore considered to be primitive. The process of biochemical evolution has been conservative, in that once the chemical mechanisms were laid down, they have not been changed (Margulis, 1982).

All organisms can be divided clearly into two groups: the simpler prokaryotes and the more complex eukaryotes. Eukaryotes are made up of cells that contain nuclei, while prokaryotes consist of cells that lack nuclei. The bacteria and their relatives make up the prokaryotes; the plants, animals, fungi, and protists make up the eukaryotes. Different activities of the eukaryotic cells are localized into special organelles. The nucleus contains the chromosomes, rod-shaped bodies that are made up of DNA, protein, and RNA. The endoplasmic reticulum, a network of membranes, is covered with ribosomes that are used for protein synthesis. Mitochondria are organelles that are involved in energy production. In plants, the eukaryotic cells have special organelles, chloroplasts, for the conversion of radiant energy from the sun to chemical energy (photosynthesis). In the prokaryotic cells there are no chromosomes but a circular strand of DNA that is not bounded by a membrane; the ribosomes are found throughout the cell; there is no endoplasmic reticulum. Prokaryotes were the earliest organisms to appear on earth. Eukaryotes appeared much later (about 2 billion years ago). The Precambrian period has been lucidly presented by Lynn Margulis in *Early Life* (1982). I am grateful to her for my discussion of early life.

The Precambrian era about which we knew very little some 25 years ago makes up about 90 percent of the earth's existence. This period of time has now been divided into Hadean, Archaean, and Proterozoic. The earth arose as part of the solar system by a condensation of dust and gases about 5 billion years ago. The basis for this conclusion rests upon work done by astronomers and astrophysicists. Rocks collected from the moon have been shown to be the same age as the earth. Space probes by man-made satellites have shown that other solar systems are being formed.

During the Hadean period (4.6 to 3.8 billion years ago) the earth, which was initially hot, cooled down sufficiently to form surface rocks as solid bodies. The Archean (3.8 to 2.6 billion years ago) was a period in which the cooled-down surface of the earth permitted the formation of the oldest earth rocks. It was during this period that chemical evolution began. From the gases that existed in the earth's atmosphere (hydrogen, methane, carbon dioxide, and water vapor) and from energy supplied by ultraviolet rays and lightening, the more complex molecules, amino acids and proteins, were formed. Later in this period prokaryotic cells appeared as well as the innovations of fermentation, photosynthesis, and nitrogen fixation. In the Proterozoic period (2.6 billion to 600 million years ago), eukaryotic cells evolved and bisexual reproduction developed. About 1 billion years ago large algae, several centimeters in diameter, had evolved.

In Ediacara, Australia, excellent impressions of Precambrian fossils were found of multicellular soft-bodied marine organisms. These resembled the skeletonized Cambrian fossils. Such Precambrian fossils have now been found in Siberia, England, and Greenland. Bacterial organisms have been found in two-billion-year-old rocks in the Gunflint iron formation of Canada. Bacterial fossils, 3.5 billion years old, were found in rocks from Australia. A curious dome-shaped rock has been found as part of the fossil record. These stromatolites consist of a mat of cyanobacteria that are capable of photosynthesis, growth and reproduction. This mat of cyanobacteria traps clay and mud particles. The sediment accumulates and the bacteria are trapped and die or move upward and grow. The mat turns into stone and thus a many layered rock, made by living organisms, is formed. Some ancient stromatolites contain fossils that resemble modern cyanobacteria. The oldest stromatolites found are 3.5 billion years old. The presence of these cyanobacteria as fossils indicates that photosynthesis occurred at that time.

The first organisms were heterotrophs (other-feeders) that fed upon the organic soup of the earth's waters. Later, organisms evolved that were capable of making their own food by means of photosynthesis. In this process, oxygen was released into the atmosphere and accumulated. It was the presence of oxygen that permitted the development of the more efficient energy-utilizing mechanism of respiration and with it the development of eukaryotic cells and finally multicellular forms. At one time oxygen was a toxic substance. New strains developed however that were oxygen-resistant. The formation of oxygen resistance may be comparable to the establishment of streptomycin resistance.

The knowledge that at one time there was no oxygen on earth is based upon the color of certain minerals that are found in rock formations. Certain substances such as iron, manganese, and uranium occur in oxidized and reduced states. Iron, for example, may be present in the oxidized state (hematite) and be rusty in color or in the reduced state (magnetite) and be darker in color. For over two billion years nearly all of the iron in rocks has been fully oxidized. This indicates that the atmosphere contained oxygen and that photosynthesis was occurring.

THE ORIGIN OF LIFE

Modern approaches to the experimental study of the origin of life were initiated in the 1930s by Oparin, a Russian biochemist. Oparin points out that the question of the appearance of life on earth is both a scientific and a religious question. The question of creation has been seriously considered by every religious system since antiquity. Today, the creation of life and the creation of the universe are under active scientific investigation.

All organisms living today arose from other living organisms (bacteria from bacteria, oak trees from oak trees, men from men). This is the Biogenetic Law—"all life arises from preexisting life." At one time it was thought that life arose continuously from non-living material. This is the idea of *spontaneous generation.* Spontaneous generation involves the transformation of inanimate material into living organisms by natural means. Special creation involves the creation of living organisms at one time from non-living material for a special purpose by a Creator. Spontaneous generation is a scientific concept that can be tested by experiments, while special creation is a religious concept that cannot be tested by science.

For many people it was clearly established that life arose from non-living material. The Chinese thought that insects arose under the influence of heat and moisture. The sacred books of India wrote that flies and beetles arose from sweat and manure. In Egypt it was thought that the soil of the Nile River gave rise to frogs, snakes, and mice. Aristotle thought that worms and snails were the products of putrefaction. An interesting recipe by Paraclesus (1500) suggested that a man could be made by placing human semen in a squash inside the belly of a horse. It was certainly the common experience of people that life arose from non-living material; weeds from barren ground, frogs from mud, mice from rags. When I worked in my grandfather's grocery store, customers would occasionally bring back a sealed box of cereal in which there were insect larvae. "How could this be," said my grandfather? It took many careful experiments performed over a period of 400 years to show that spontaneous generation does not occur today.

It is curious that while we have looked upon the creation of living organisms from inanimate material as a divine achievement, we have at the same time viewed this type of creation as commonplace.

Although life arises from preexisting life today, how did life get started "in the beginning"? Where did the original organisms come from? Various biologists have recognized that spontaneous generation remains a logical necessity. In the beginning there could have been divine intervention. Maybe life arose through supernatural events. While this view may be completely satisfying to theology, it is useless science. Science can only explain natural events. And so, from 1860, after Pasteur's experiments, until 1930, no experiments were done because no one knew just what to do. In the 1930s some ideas by the Russian biochemist, Oparin, opened up the field again to investigation.

The major changes in thinking that stimulated investigations were that organic compounds could be made abiotically, as well as by organisms; that there did not occur a chance coming together of simple elements to form a complex living unit all at once, but rather a gradual increase in complexity from elements, to inorganic molecules, to organic molecules, to complex biochemical substances occurring over a period of three billion years; and that living things have certain molecular combinations and life therefore arises as a consequence of organization rather than because of certain mysterious properties or "vital forces."

Oparin viewed the evolution of matter from the simple inorganic to the complex organic as steps in the unfolding of a single process. Oparin proposed that the primitive atmosphere of the earth contained methane (CH_3), ammonia (NH_3), water vapor (H_2O), hydrogen (H_2) but no oxygen. He suggested that in this reducing atmosphere and under the force of ultraviolet radiation and other forces of energy that organic compounds would be made. He thought that the primitive oceans contained so much organic matter that they constituted a hot dilute soup. Experimental proof for the abiotic synthesis of organic compounds from the substances suggested by Oparin was obtained in 1953 by Stanley Miller. Miller circulated a reaction mixture of water vapor, methane, ammonia, and hydrogen continuously for one week over an electric spark. Analysis showed that the water trapped in the bottom of the container contained a mixture of amino acids. This experiment opened up a field of study and led to the acceptance of the idea that organic compounds can be formed abiotically. By use of different reaction mixtures that included carbon monoxide (CO), nitrogen (N_2), carbon dioxide (CO_2), cyanide (HCN), and which used electric sparking, bright sunlight, heat, ultraviolet rays, and high-energy rays, it has been possible to synthesize amino acids, small peptides, polypeptides, nucleotides, adenosine triphosphate, sugars, and polysaccharides. Although these experiments show that many of the organic compounds found in living organisms today could have been formed abiotically, they do not prove that this is the way it happened three billion years ago.

Organic compounds are probably widely distributed throughout the universe and conditions may exist whereby these compounds could be synthesized abiotically. There is now a search for life and for organic compounds in other planets and in the universe. This is called "exobiology" and is part of the space program. The temperatures and probable atmospheric gases of the planets in our solar system have been determined. Meteorites furnish direct evidence for the chemical composition of extraterrestrial bodies and contain lipids, purines, pyrimidines, and amino acids. Some meteorites have bodies which look like microbes. Thus, the origin of life appears to be a natural event. Given the original chemicals and the physical conditions on earth, life was inevitable. The unmanned space probes show that new stars and solar systems are being made. Other solar systems may have the means to support life. Astronomers think that there are many planets in the solar systems that could support life. Given the appropriate chemical and physical conditions, life is inevitable.

CHANCE AND DETERMINISM

George Wald (1954) has considered the probability of improbable events such as the making of a functional protein as follows. With every event we can assume a probability. If we toss a penny, the chances of obtaining "heads" is $\frac{1}{2}$ and of obtaining "tails" is $\frac{1}{2}$. When we toss two dice, the probability of obtaining a "12" is $1/36$. If we toss pennies or dice many times, we will be able to demonstrate the predicted probabilities. In dealing with the origin of a protein, we are dealing with an event that needs to happen only once. Under these conditions, an improbable event becomes a certainty, if there are enough trials. If the number of trials is sufficient, we can achieve any degree of certainty. What is impossible in a lifetime becomes a virtual certainty over 3 billion years. "Time is the hero." If the chances of getting "heads" from one toss of a penny are $\frac{1}{2}$, then the chances of getting 10 tails in 10 tosses are $(\frac{1}{2})^{10} = 1/1{,}024$. Thus, the chances of getting at least 1 heads in 10 tosses are $1{,}023/1{,}024$. Suppose that an event occurs one time in a thousand $(1/1{,}000)$. In 1,000 trials, the chances that the event will not occur are $(999/1{,}000)^{1{,}000} = 37/100$; and therefore the chances that it will occur at least once in 1,000 trials are $63/100$. In 10,000 trials, the chances that a $1/1{,}000$ event will occur at least once are $9{,}999/10{,}000$.

Some states sell lottery tickets, and an investment of one dollar may earn one million dollars. People who buy these tickets know that someone will win, but they cannot predict who will win. To become a millionaire depends on an event that need happen only once. In the sense that someone in the population will win, the lottery is deterministic. In the sense that a particular person will win, the lottery is a matter of chance.

If life is a cosmic event, if it is inevitable, are we dealing with chance or determinism? Suppose that the whole business were to start over again, would we end up with life, would we end up with insects, whales, and people, or with presently unknown forms of life. Would we end up with you and me?

> Thus, we are not alone in the Universe. Life is a cosmic event. On this planet which is our home, we have every reason to wish it well. Yet, should we fail, all is not lost. Our kind will try again elsewhere. (George Wald, 1954).

THEOLOGICAL AND SCIENTIFIC EXPLANATION

Over the years, people have made observations of events of the world about them. Early in their civilization, it was customary to relate these events to supernatural causes. With the development of science we find that many events that were explained by the whim of a deity are now explained by natural causes. The phenomena of floods, lightening, and thunder can be explained by physical laws. They are not a reaction to the sinfulness of man. Epidemics can be explained by microorganisms. The sun and moon are no longer deified. The rotation of the planets around the sun can be explained by natural laws. Today, the age of the earth, the origin of life, the meaning of fossils, and the evolution and diversity of species have natural explanations. They have been removed from the realm of theology to science. The origin of life is now a scientific rather than a scriptural matter. There is agreement among scientists that the earth is quite old (4.5 billion years); that a chemical evolution preceded a biological evolution for millions of years. Whatever the answer to the origin of life, it will turn out to be an explanation based on the laws of chemistry and physics. It is not a six-day miracle, but rather a 4.5-billion-year miracle.

SOME CREATION STORIES

The Babylonian View

Most stories of the origin of man and of the Deluge depend upon supernatural causes. Every civilization has proposed a Creation Story and/or a Flood. There are hundreds of such stories. There is a famous Babylonian story describing the sending of a flood by the gods. Atrahasis (the Babylonian Noah) and his family was commanded to build a large ship and to take the birds and the beasts of the field onto the large ship. Although the story of the actual flood was not preserved in the available Atrahasis tablets, it is found also in the story of Gilgamesh (an early hero of Mesopotamia). In this story Ut-napishtim is the builder of the ark and he survived seven days of flood and wind. His ark was grounded on Mount Nisir. Ut-napishtim sent forth a dove, a swallow, and then a raven before the earth reappeared after the waters receded.

The Traditional Chinese View

The world was created by a giant of cosmic proportions named Pan K'u, who labored for enormous periods of time to build the world and construct the four great pillars which support the sky. At the end of his labor this huge man died. In his death and disintegration the universe came into being. His eyes became the sun and moon. The muscles and sinews of his body became the earth and mountains, his bones the minerals of the soil. The hair on his body formed the basis of trees and other planets, while his veins formed the courses of great rivers and his blood the water of the rivers and seas. And finally the insects which inhabited his living body, the fleas and lice which are constant companions of most of mankind, became human beings.

View of the Havasupai of the Grand Canyon

Before there were any people on earth there were two gods, Tochopa of goodness and Hokomata of evil. Tochopa had a daughter named Pu-keh-eh, whom he hoped would become the mother of all living things. Hokomata the evil one was determined that this would not take place, and he covered the world with a great flood. Tochopa, the good, felled a great tree and hollowed out the trunk. He placed Pu-keh-eh in the hollow trunk, and when the water rose and flooded the earth she was secure.

Finally, the flood waters fell and mountain peaks emerged. Rivers were created and one of them cut the great gushing fissure which became the Grand Canyon.

Pu-keh-eh in her log came to rest on the new earth. She stepped forth and beheld an empty world.

When the land became dry, a great golden sun rose in the east and warmed the earth and caused her to conceive. In time, she gave birth to a male child. Later, a waterfall caused her to conceive and she gave birth to a girl. From the union of these two mortal children came all the people on the earth. The first were the Havasupai, and the voice of Tochopa spoke to them and told them to live forever in peace in their canyon of good earth and pure water where there would always be plenty for all.

A SUGGESTION. You may find it interesting to read some of the hundreds of creation stories and to compare them.

CHAPTER 14

VIEWING THE BIBLE

FAITH, like science, should not be blind but, indeed, demands the honesty of scholarship. A considerable effort has been made by biblical scholars over the past 1,500 years in biblical analysis. These analyses have involved translation and interpretation from ancient Hebrew and Greek texts. It is clear to these scholars that every translation necessitates interpretation, because the words in one language do not have exactly the same meaning in another language and because the meaning of words even in the same language may differ in time. It is clear to many scholars that much of the Bible is written as allegory and poetry and therefore should not be taken literally. How many English versions of the Bible have been written? Are they the same in wordage? Do they all say the same thing? Read Psalm 23 and the Lord's Prayer in five different versions of the Bible. Where did the Bible come from? Are they the same in content? Are there books in the Bible of one sect that are not in the Bible of another sect? How was it decided just what should go into the Bible?

Creation

The Bible starts with a story of Creation. What is this story (Genesis 1:1-2:25)? Notice that there are two accounts of creation by God. Biblical scholars think that they were written at different times. The first Genesis creation story (the "Priestly" or "P" account) has been shown to have a completely different style and vocabulary from the second creation story (the "Yahwist" or "J" account). The "J" account is believed to have been written about 950 B.C. and the "P" account about 400-500 B.C.

ACCOUNT 1 (Genesis 1:1-2-4a):

In the beginning God created the heavens and the earth. The earth was without form and void, and darkness was upon the face of the deep. How long was the earth "without form and void" before God created light? Is "the beginning" the same as "day one"? What is meant by "face of the deep"?

Day One

God made light and divided the light from the darkness. He called the light Day and the darkness Night. There was evening and there was morning on day one.

Day Two

God made a firmament in the midst of the waters. He separated the waters that were above the firmament from the waters that were below the firmament. God called the firmament, Heaven. What is a "firmament"? What is the meaning of "the waters above and the waters below"?

Day Three

God gathered the waters into one place; He called them Seas. God let dry land appear; He called it Earth.

God bade the Earth to bring forth different kinds of vegetation, including fruit trees and other seed-bearing plants.

Day Four

God made two great lights in the Heavens to give light upon the earth. The greater light ruled over day and the lesser light ruled over night. He made the stars. Notice that God made light and darkness on day one, but He did not make the sun and moon until day four. Where did the light on the first three days come from? He created vegetation before he created the sun. He created the earth before He created the sun.

Day Five

God bade the waters bring forth all kinds of living creatures. He created birds. He created great sea monsters. God created the whales.

Day Six

God bade the Earth to bring forth land animals. God created man in His image; male and female He created them.

Day Seven

God rested from all of the work which He had done in creation.

ACCOUNT 2 (2:4b-25):

Before there was even any vegetation or animals on earth, God created man. God then planted a garden in Eden. God created every kind of animal and brought them to man to name. God then made a woman from one of man's ribs.

Notice that in Account 1 that man was created after the creation of the plants and animals; in Account 2, man was created first, then trees, then animals, then woman.

How does the sequence of creations in the Bible compare with the sequence as proposed by biologists? There is a difference in time, from a period of six days to millions of years. In the biblical version the water creatures and the birds were created before the land creatures. Seed-bearing trees were created before animals. In the biological version, reptiles were created before water-dwelling mammals such as whales, and fishes existed before fruit trees.

Since creation was completed within the short period of six days, at one time all "kinds" of organisms lived at the same time. People and dinosaurs must have been contemporaries. Their remains, therefore, should be found together in the same strata of the earth's surface. The different strata of rock making up the earth's crust, however, clearly show that a sequence of different kinds of organisms existed at different times. Fossils of people do not appear in the same strata as dinosaurs, despite the claims at Glen Rose, Texas.

All of the creation events as described in the Bible are due to supernatural causes, to miracles. They cannot be explained by science. Many people believe that science dare not even ask or question. The Bible is not required to present evidence for its conclusions. Biologists, however, who attempt to explain the "creation" must look for natural causes. They are obliged to present evidence for their conclusions. The scientific explanation is always open to challenge and modification. To many people, the religious explanation seems to be true for all time and cannot be challenged. This is why the Bible is a book of religion and not a book of science. It is important to us all that the Bible remain as our book of religion. It is not written to be a scientific text.

On Biblical Literalism

From the Holy Bible Catholic Version. The Douay Version of the Old Testament. First published by the English College at Douay 1609 A.D. P.J. Kennedy and Sons, New York, 1950.

The Bible was written by men who were inspired by God. Their writings are the authority of God. The Bible is never wrong.

"The meaning of the Bible is not always easily ascertained. Truth is affirmed in one way in poetry, in another in prose; in one way in an allegory, in another in a history; in one way history which aims at exact statement; in another, history which admits of a certain freedom in narration. Interpretation is difficult of any ancient literature and is particularly great in the case of ancient books of the Bible.

"No dates can be assigned to events narrated in the first eleven chapters of Genesis (Creation to Abraham). The period is certainly much longer than was thought, 2023 years. One would have to admit a manifest conflict between the Bible and science."

From the Holy Bible (Revised Standard Edition). Thomas Nelson and Sons, New York, 1952.

"The first English version of the Scriptures made by direct translation from the original Hebrew and Greek, and the first to be printed, was the work of William Tyndale. He met bitter opposition. He was accused of willfully perverting the meaning of the Scriptures, and his New Testaments were ordered to be burned as "untrue translations." He was finally betrayed into the hands of his enemies, and in October 1536, was publically executed and burned at the stake."

"Yet Tyndale's work became the foundation of subsequent English versions." "The translators who made the King James Version took into account all of the preceding versions." "It owed most, especially in the New Testament, to Tyndale."

"Yet the King James Version has grave defects. By the middle of the nineteenth century, the development of biblical studies and the discovery of many manuscripts, more ancient than those upon which the King James Version was based, made it manifest that these defects are so many and so serious as to call for revision of the English translation." These revisions in the nineteenth and twentieth centuries however have stayed close to Tyndale-King James tradition (from Preface).

From the Interpreter's Bible, A Commentary in Twelve Volumes. Abingdon Press, New York, 1952.

These books deal with an analysis of the Old and New Testaments. There were 160 biblical scholars involved in this study. The question is raised (page 460) as to whether the early figures in Genesis are part of a real history or products of the imagination. The view is taken that Genesis begins as myth. Few people will suppose that Adam and Eve and the

Garden of Eden belong to factual history. Cain, Abel, Methusella, and Noah came down to us as legends. There are questions whether there was an actual Abraham, Isaac, or Jacob.

The Bible as a Book of Science and a Book of History

At one time the Bible was considered to be an accurate history of the earth and the origin of man. Studies in science and biblical interpretation have challenged this biblical view. According to the Bible, creation occurred in six days, the Earth was the center of everything, and the sun and moon rose and set over a flat earth. These views were challenged by Copernicus and Galileo who showed that the earth is not the center of the universe but a little satellite that goes round the sun. Geologists and paleontologists showed that the earth and its organisms were not made in six days but slowly formed over millions of years. Charles Darwin proposed, and succeeding biologists have demonstrated, that present-day organisms have evolved from other kinds of organisms and that man arose from organisms that were not man.

You may find it interesting to read the story of Noah and the Ark and to critique it from the standpoint of scholarship rather than of faith. What questions occur to you concerning this account that call for further explanation? A group of people are currently looking for Noah's Ark on Mount Ararat, Turkey. Suppose that they find a boat on Mount Ararat, how can they establish that it is the ARK?

Despite the fact the certain parts of the Bible appear to be literally false, the Judeo-Christian tradition is strong in the minds of men. Christianity did not collapse with the discovery of planetary rotation. It did not collapse with the general acceptance of evolution. Other religions, each with their own legends and traditions, also persist as important institutions in society.

Scholarship in both religion and science deals with a search for meanings. This search may involve the study of the universe around us by scientific methods or the study of our goals and of our purpose on this planet. In neither of these most human of organismal enterprises is our quest completed. We need to preserve our right to question and to think in both the areas of science and religion. We need to preserve an intellectual environment in which our descendents have the right to prove us wrong. Scholarship has the same ultimate goals in both the fields of science and religion.

Scientific methods and attitudes have been used not only in the biological and physical sciences but also in the study of religion. Books have been written that arise from such studies of religion and ethics. In these studies, religious concepts are challenged and theories are tested. Scholars who carry on this research think that this method of criticism contributes to a stronger religious faith and moral idealism.

Sacred Writings (After Toy)

Most religions have sacred writings that embody their beliefs. These sacred writings become canon and contrary views are considered to be heretical. The canon of one religion is heretical to that of another religion. The selection of sacred books is decided by councils of men who decide by a majority vote which writings shall be chosen. The decisions are based on the political climate and national literature at the time and on the need to protect a religious group against rivalry. One finds, therefore, that different sects select different writings.

The Palestinian Jews, about 1,000 A.D., selected their canon (and after hot debate) in the form that the Hebrew Old Testament now appears. The Alexandrian Jews accepted the Palestinian canon but added a group of books (the Apochrypha) that the Palestinians rejected. Early Christianity adopted the Jewish canons and in addition produced its own literature. During this period there was doubt about certain books, and therefore a definite standard was set toward the end of the fifth century by Pope Gelasius. The Apochrypha was used by early Christian writers because it was found in the Greek translation of the Old Testament which was the Old Testament text used by the Christians. Their popularity was so great that Jerome, against his judgment, included them in his translation (the Latin Vulgate). The Council of Trent (1546) accepted them as canon, and they are still so regarded by the Roman Church. The Apochrypha was included in the original King James Version but some Protestant sects do not include the Apochrypha.

When certain sacred books are considered to be infallible, all of its statements (including those relating to science) are thought to be literally true. If interpretation is necessary because of new secular concepts, the image of infallibility is maintained. Religious bodies tend to claim absolute authority in religious matters, but they tend to define their domain to include secular things. The belief that a sacred book is final authority often results in the limitation of freedom of thought, since certain matters are excluded from discussion.

On Untouchable Truths

Are there certain ideas that are so crucial to the survival of a society that they should not be questioned? In the field of science one is obliged to question every concept. In religion, one is limited in what one dare question. "You may not question this, but you may question that." "My church tells me that I must accept the idea of creationism, but I am permitted to believe whatever I want about the rotation of the planets or atomic theory." "This concept is an eternal truth put forth by God, you may not challenge it." Let us consider this matter of "untouchable truths."

In every religion, in every country, there are rallying cries or creeds that are central to the purpose and conduct of that society.

"We hold these truths to be self-evident, that all men are created equal." This is a statement of the ideals and goals of the United States. Based upon their experiences in other countries where human equality in opportunity, religion, and law did not exist, the colonists decided to form a country based upon this creed. One does not ask whether it is true or false. It is an untouchable ground rule.

Hebrews proclaim, "Hear O Israel the Lord is God the Lord is One." This is a creed to rally the moral conduct and the purposes of the Hebrew religion. It is a reaction to societies that practice idolatry.

The Ten Commandments are a set of rules that a society lives by because they keep the society together. The validity of the Ten Commandments is not dependent upon whether or not they came from Mount Sinai. They are ground rules for people to live together.

People do not live as hermits. They interact with other people. The welfare of the individual is enhanced by his associations with others. This experience in group living places certain restrictions upon him. The perpetuation of the society often requires individual sacrifice. This sacrifice may entail not only death on the battlefield but also a right to the freedom of thought. This is shown in an extreme form in present-day Iraq where people are told what to think and what to believe. They are not permitted to question or to challenge or to innovate. The country is controlled by a fundamentalist sect of Shiite Moslems. The society is perpetuated but at the cost of individual intellectual freedom.

The Garden of Eden and Noah's Ark are untouchable truths for biblical literalists. They are forced upon us for our own good. Can one be a Jew or a Christian and question these truths? Would Christianity collapse without the Garden of Eden? The question is not whether untouchable truths should be challenged by society. The fact is that they are challenged. At one time everyone believed the creation story as given in the Bible. Today, this story is considered by many Christians and Jews to be myth. The creation story was an untouchable truth of the eighteenth century but not of the twentieth century.

If one examines the history of religious thought over the past 2,000 years, one finds that changes in untouchable truths have occurred. Moreover, these changes have been brought about by religious people with a deep sense of devotion to God. Consider the major differences between Judaism and Christianity, Catholicism and Lutheranism, among the many Protestant sects. Consider the differences among Christianity, Islam, and Buddhism. Each religious group can be described by its own array of untouchable truths.

Christianity and Judaism are strong and viable institutions today because they serve a psychological and emotional need. They help each of us to face an uncertain future and to face tragedy. The religion gives a certain stability to the society by reminding it of its spiritual goals. Its existence depends upon its usefulness to us today and not to a literal Garden of Eden. It serves no purpose to ask whether Noah's Ark is literally true or false. It is demeaning (and bad strategy) to try to justify the worth of a religion on the basis of the historic truth of an unprovable myth or by resorting to threats and promises.

Each religion needs to be relevant to its period in history. It needs to accommodate to new scientific ideas, new political concepts, new historical experiences, new interactions among migrating peoples. A twentieth century religion needs to be different than a religion of the fifteenth century. A religion that is unable to adapt to a new intellectual environment will not survive. An unchanging religion can only survive in an intellectually static world.

The truths and rallying cries of one century are not necessarily relevant to another century any more than the rallying cries of one country are relevant to another. We cannot hold all truths to be self-evident for all time. Each untouchable truth is temporal in scope. No one can anticipate what civilization will be like 1,000 years from now. Those ideas that are crucial for our existence today may not be relevant to the year 3,000. It is a sufficient challenge and a worthy goal to make the religion relevant to the times. The religion of one period needs to frame a set of goals and attitudes upon which the next generation will build rather than choke.

CHAPTER 15

A MATTER OF ATTITUDE

LAST YEAR I visited the National Museum of Air and Space located in Washington, D.C. The exhibits show the history and achievements of man in space. I thought to myself that all of this could not have been possible without the contributions of Copernicus, Galileo, and Newton or without the freeing of science from biblical literalism. Before we had set out to the moon in 1969, we already knew the distance and the length of time needed to get there. We knew exactly the amount of the gravitational force on the moon even though we had never been there. But this museum is more than merely cold history and cold science. This is shown by two huge murals that hang in the main lobby. One of these shows the Grand Canyon, magnificant in its size, and with a small airplane flying above it. The other mural shows the first men on the moon and with the planet earth in the distance. Under these murals was the following, "Billions of years ago in a burst of energy the planets, the stars, and all of the infinite wonders of the Universe were born." And, "Over the lunar surface is suspended the giant sphere of the planet earth; above it the sun casts its radiance upward to the galaxies and a new destiny for Mankind."

The museum is hallowed ground. It glorifies man's flight to the moon by use of art and poetry. "The infinite wonders of the Universe" expresses an appreciation for the vastness of time and space. It may be a universe that we will never completely understand. But, we will try and we will discover other wonders to challenge our minds and uplift our spirits. It is not sinful to stand up straight and look the Universe in the eye and ask questions. It is discouraging to think that we dare not ask or that we will never know. The natural philosophers of 300 years ago were involved in a description of the earth, its creatures, the planets, the stars, and in a search for the laws of nature. They considered their efforts to be noble work because they were investigating God's world. "A new destiny for mankind" expresses a hope that the future will be better for man than the past. It expresses a conviction that man has the ability to improve the human condition by his own efforts.

I think back to the woman whose daughter was frightened and upset by her biology teacher and the idea of evolution. What kind of child are we rearing? Should we allow her to attend public school? Do we let her go to museums? Do we let her read newspapers or magazines? Do we allow her to listen to lectures at the Grand Canyon? Is she permitted to read books on astronomy, or geology, or oceanography? What does her Sunday School teacher answer when she states, "My science teacher says that the earth is 4 billion years old," or birds are descended from reptiles, or when she finds that other students in class manage to attend church and accept the concept of evolution? Should the teacher answer by saying that scientists are atheists, or evil or dishonest, or stupid, or badly motivated and insincere? Is the answer to create a distrust of science or to say that science deals with matters that are inconsequential? How do we prepare a generation of children to face a world with diverse people and contrary ideas. Should Congress pass laws to close the Museums of Natural History? Should we close the Dinosaur National Park?

At this time in history, people are able to ask questions that will have a great impact on both science and religion. Is there a biological basis for human behavior? Can the characteristics of man be modified by the insertion of genes from other species? How large is the Universe and how old is it? Is there life in other solar systems? Can we synthetically make living organisms? How small is the smallest particle? The twenty-first century will be exciting and amazing for scientific discovery. It will have a considerable effect on religious beliefs. Do you have a religion that can cope with new ideas?

A student told me that the Bible has in it such amazing stories that it must be believed. The creation within six days of an entire universe and an earth with plants and animals and man is certainly a miraculous achievement. It inspires reverence and awe. But the study of evolution is also amazing. Think of a universe that is 20 billion years old and of infinite size with an increase in the complexity of substances from atoms to large molecules, from simpler organisms to more complex organisms. Think of mountains being, at one time, under the sea, of South America and Africa joined together. And think of the achievements of a species, *Homo sapiens,* that is able to learn these things because it has the ability and the right to challenge old ideas and to innovate.

Last year when I visited the Grand Canyon, I took a trip with the Ranger Naturalist to search for fossils. We found rocks that contained fossil sponges, brachiopods, and sea lilies. The rocks on which we stood were 200 million years old. The land at one time was under the sea. It is a wonderful experience to stand on these rocks and look out at the Grand Canyon. Many people find this site to be inspirational and respond by holding church services here. There is a cross on the edge of the canyon. We have a poignant scene of a naturalist who believes he is standing on rocks that are 200 million years old, looking at a canyon that is 6 million years old and at the same time praying to a Creator.

It was with reverence and awe that Charles Darwin viewed the idea of the creation of organisms by means of the process of evolution. The last sentences of *On the Origin of Species* show this sense of awe and inspiration. "There is grandeur in this view of life, with its several powers, having been originally breathed by the Creator into a few forms or into one; and that, whilst this planet has gone cycling on according to the fixed law of gravity, from so simple a beginning endless forms most beautiful and most wonderful have been and are being evolved."

A Final Challenge

In the 1950s it was customary for many universities to hold a religious emphasis week. One of the seminars in which I participated addressed the question, "In an age of science, what is one to believe about the relation of man to God?" Sit down and answer this question.

BIBLIOGRAPHY

Ancestors, Four Million Years of Humanity. New York, American Museum of Natural History, 1984.

Ayala, Francisco, J.: *The Origin of Species.* Elon College, Carolina Biological Supply, 1983.

Ayala, Francisco, J. and Valentine, James W.: *Evolving, The Theory and Processes of Organic Evolution.* Menlo Park, Benjamin Cummings, 1979.

Barton, George A.: *The Religions of the World.* 3rd ed. Chicago, University of Chicago Press, 1929.

Bentley, Eric: *The Recantation of Galileo Galilei.* New York, Harper and Row, 1972.

Bixby, William: *The Universe of Galileo and Newton.* New York, American Heritage, 1964.

Bodmer, Walter F. and Cavalli-Sforza, Luigi L.: *Genetics, Evolution and Man.* San Francisco, Freeman, 1976.

British Museum (Natural History). *Origin of Species.* New York, Cambridge, 1981.

Bronowski, Jacob: *Magic, Science and Civilization.* New York, Columbia, 1952.

Bronowski, Jacob and Mazlich, Bruce: *The Western Intellectual Tradition.* New York, Harper and Row, 1962.

Campbell, Joseph: *Myths To Live By.* New York, Viking, 1972.

Carson, Hampton L.: Hawaii, a showcase of evolution. Natural History *91, No. 12,* 1982.

Clark, W. E. LeGros: *The Fossil Evidence for Human Evolution.* Chicago, University of Chicago Press, 1955.

Clark, W. E. LeGros: *Man-Apes or Ape-Men?* New York, Holt, Rinehart and Winston, 1967.

Colbert, Edward H.: *Wandering Lands and Animals.* New York, Dutton, 1973.

Colbert, Edward H.: *The Evolution of the Vertebrates.* 3rd ed. New York, Wiley, 1980.

Crompton, A.W. and Parker, Pamela: Evolution of the mammalian masticatory apparatus. *American Scientist, 66:*192, 1978.

Darwin, Charles: *On the Origin of Species* (1859). New York, Atherton, 1967.

DeChardin, Pierre Teilhard: *The Appearance of Man.* New York, Harper and Row, 1956.

Dobzhansky, Theodosius: *Mankind Evolving: The Evolution of the Human Species.* New Haven, Yale, 1962.

Dodson, Edward O. and Dodson, Peter: *Evolution, Process and Product.* New York, Van Nostrand, 1976.

Dubos, Rene and Dubos, J.: *The White Plague: Tuberculosis, Man and Society.* Boston, Little, Brown, 1952.

Eckhardt, Robert B.: *The Study of Human Evolution.* New York, McGraw-Hill, 1979.

Eiseley, Loren: *Darwin's Century.* Garden City, Doubleday, 1961.

Fenman, Frank: *Myxomatosis in Australian Wild Rabbits—Evolutionary Changes in an Infectious Disease.* The Harvey Lectures, New York, Academic, 1957.

Ford, Edmund B.: *Mendelism and Evolution.* London, Methuen, 1960.

Frieden, Earl: The chemical elements of life. *Scientific American, 227, No. 1:*52, 1972.

Futuyma, Douglas J.: *Science on Trial.* New York, Pantheon, 1983.

Gabriel, Mordecai L. and Fogel, Seymour: *Great Experiments in Biology.* Englewood Cliffs, Prentice-Hall, 1955.

Gallant, Roy A.: *How Life Began, Creation Vs. Evolution.* New York, Four Winds, 1975.

Glaessner, Martin F.: Precambrian fossils. *Scientific American, 204, No. 3:*72, 1961.

Gould, Stephen J.: *Hen's Teeth and Horse's Toes.* New York, Norton, 1983.

Greene, John C.: *The Death of Adam.* Ames, Iowa State, 1959.

Hardin, Garrett: *Nature and Man's Fate.* New York, New American, 1959.

Hartl, Daniel L.: *A Primer of Population Genetics.* Sunderland, Sinauer, 1981.

Holy Bible, Revised Standard Version. New York, Thomas Nelson, 1952.

Holy Bible, Catholic Version. New York, Kennedy, 1950.

Hudson, P.S. and Richens, R.H.: *The New Genetics in the Soviet Union.* Cambridge, Imperial Bureau of Plant Breeding, 1946.

Hurley, Patrick M.: The confirmation of continental drifts. *Scientific American, 218, No. 4:*52, 1968.

Huxley, Julian: *Heredity, East and West.* New York, Schuman, 1949.

The Interpreter's Bible, A Commentary in Twelve Volumes. Vol. 1, New York, Abington, 1952.

Johanson, Donald C. and Edey, Maitland A.: *Lucy, The Beginnings of Mankind.* New York, Simon and Schuster, 1981.

Keosian, John: *The Origin of Life.* New York, Reinhold, 1968.

Kerr, Richard A.: Precisely measuring the past million years. *Science, 221,* 1041, 1983.

Kettlewell, H. Bernard: Darwin's missing evidence. *Scientific American, 200,* March: 48, 1959.

King, Mary C. and Wilson, A.C.: Evolution at two levels in humans and chimpanzees. *Science, 188:*107, 1981.

Kollan, E.J. and Fisher, C.: Tooth induction in chick epithelium: expression of quiescent genes for enamel synthesis. *Science, 207:*993, 1980.

Kurten, Bjorn: Continental drift and evolution. *Scientific American, 220, No. 3:*54, 1969.

Langford, Jerome L.: *Galileo, Science and the Church.* Ann Arbor, University of Michigan, 1966.

Leakey, Richard E. and Lewin, Roger: *People of the Lake.* Garden City, Anchor, 1978.

Lerner, I. Michael and Libby, William J.: *Heredity, Evolution and Society.* San Francisco, Freeman, 1976.

Maclagan, David: *Creation Myths.* New York, Thames and Hudson, 1977.

Margulis, Lynn: *Early Life.* New York, Van Nostrand Reinhold, 1982.

Medvedev, Zhores: *The Rise and Fall of T.D. Lysenko.* New York, Columbia, 1971.

Miller, Stanley, L.: A production of amino acids under possible primitive earth conditions. *Science, 117:*528, 1953.

Minkoff, Eli C.: *Evolutionary Biology.* Reading, Addison-Wesley, 1983.

Moore, John A.: Science as a way of knowing—evolutionary biology. *American Zoologist, 23:*1, 1983.

Morris, Henry M.: *The Remarkable Birth of the Planet Earth.* San Diego, Creation Life, 1972.

Murray, N. Patrick and Buffaloe, Neal D.: *Creationism and Evolution, The Real Issues.* Little Rock, The Bookmark, 1981.

Neil William: *The Plain Man Looks at the Bible.* London, Fontana, 1957.

Oparin, Aleksandr I.: *The Origin of Life.* 2nd ed. New York, Dover, 1953.

Pai, Anna C.: *Foundations of Genetics.* New York, McGraw-Hill, 1974.

Prenant, Marcel: *Biology and Marxism.* New York, International, 1943.

Raff Rudolph A. and Kaufman, Thomas C.: *Embryos, Genes and Evolution.* New York, MacMillan, 1983.

Raup, David M. and Stanley, Steven M.: *Principles of Paleontology.* San Francisco, Freeman, 1978.

Ruse, Michael: *The Darwinism Revolution.* Chicago, University of Chicago Press, 1979.

Sears, Paul B.: *Charles Darwin, The Naturalist as a Cultural Force.* New York, Scribner's, 1950.

Shorrocks, Bryan: *The Genesis of Diversity.* Baltimore, University Park, 1979.

Simpson, George G.: *The Meaning of Evolution.* New Haven, Yale, 1949.

Simpson, George G. and Beck, William S.: *Life, and Introduction to Biology.* 2nd ed. New York, Harcourt Brace, and World, 1965.

Singer, Charles: *A Short History of Science.* London, Oxford, 1941.

Sinnott, Edmund W., Dunn, Leslie, C., and Dobzhansky, Theodosius: *Principles of Genetics.* 5th ed. New York, McGraw-Hill, 1958.

The Situation in Biological Science. *Proceedings of the Lenin Academy of Agricultural Sciences of the U.S.S.R.* New York, International, 1949.

Sonneborn, Tracey: Heredity, environment and politics. *Science, 111:*529, 1950.

Stanfield, William D.: *The Science of Evolution.* New York, MacMillan, 1977.

Stebbins, G. Ledyard: *Darwin to DNA, Molecules to Humanity.* San Francisco, Freeman, 1982.

Stine, Gerald J.: *Biosocial Genetics.* New York, MacMillan, 1977.

Strickberger, Monroe W.: *Genetics.* New York, Macmillan, 1976.

Toy, Crawford H.: *Introduction to the History of Religion.* Cambridge, Harvard, 1924.

Volpe, E. Peter: *Understanding Evolution.* 4th ed. Dubuque, Brown, 1981.

Wald, George: The origin of life. *Scientific American, 191, No. 2:*44, 1954.

Wilson, Edward O.: *On Human Nature.* Cambridge, Harvard, 1978.

Yunis, Jorge J., Sawyer, Jeffrey P. and Dunham, Kelly. The striking resemblance of high resolution G-banded chromosomes of man and chimpanzee. *Science, 208:*1145, 1980.

Yunis, Jorge J., and Prakash, Om: The origin of man: a pictorial legacy. *Science, 215:*1529, 1982.

INDEX